# DESIGNING MULTILINGUAL EXPERIENCES IN TECHNICAL COMMUNICATION

# DESIGNING MULTILINGUAL EXPERIENCES IN TECHNICAL COMMUNICATION

LAURA GONZALES

UTAH STATE UNIVERSITY PRESS
*Logan*

Published by Utah State University Press
An imprint of University Press of Colorado
245 Century Circle, Suite 202
Louisville, Colorado 80027


Printed in the United States of America

The University Press of Colorado is a proud member of the Association of University Presses.

The University Press of Colorado is a cooperative publishing enterprise supported, in part, by Adams State University, Colorado State University, Fort Lewis College, Metropolitan State University of Denver, University of Alaska Fairbanks, University of Colorado, University of Northern Colorado, University of Wyoming, Utah State University, and Western Colorado University.

∞ This paper meets the requirements of the ANSI/NISO Z39.48-1992 (Permanence of Paper).

ISBN: 978-1-64642-275-3 (paperback)
ISBN: 978-1-64642-276-0 (ebook)
https://doi.org/10.7330/9781646422760

Library of Congress Cataloging-in-Publication Data

Names: Gonzales, Laura, author.
Title: Designing multilingual experiences in technical communication / Laura Gonzales.
Description: Logan : Utah State University Press, [2022] | Includes bibliographical references and index.
Identifiers: LCCN 2022019662 (print) | LCCN 2022019663 (ebook) | ISBN 9781646422753 (paperback) | ISBN 9781646422760 (ebook)
Subjects: LCSH: Communication of technical information—Case studies. | Multilingual communication—Case studies. | Technology—Translating. | Technical writing. | Intercultural communication.
Classification: LCC T11 .G64 2022 (print) | LCC T11 (ebook) | DDC 601/.4—dc23/eng/20220613
LC record available at https://lccn.loc.gov/2022019662
LC ebook record available at https://lccn.loc.gov/2022019663

Publication of this text was supported by the University of Florida College of Liberal Arts and Sciences and Center for the Humanities and the Public Sphere (Rothman Endowment).

Original cover illustration by Elena García Ortega, CEPIADET A.C.

*Para mi mamita hermosa,*
*Fabiola Margitt Inmaculada*
*Joffre Bergann de Gonzales*

# CONTENTS

# ACKNOWLEDGMENTS

This book would not have been possible without the collaboration, guidance, and labor of so many people. First, I am grateful to the multiple collaborators who shaped this work, including the Centro Profesional Indígena de Asesoría, Defensa, y Traducción (CEPIADET) and the friends and collaborators I gained in that space, including Abigail Castellanos García, Tomás López Sarabia, Edith Matías, Elena Ortega, Gaby De León, Shara Huaman Julluni, Nora Rivera, Mónica Morales-Good, and Cristina Kleinert. I am also indebted to Erika Hernández Cuevas and her work with Intérpretes y Promotores Interculturales for support, encouragement, translation work, and advocacy. Thank you especially to Nora Rivera for introducing me to this amazing group of people, for your feedback and guidance, and for your valued friendship.

I also thank my partners at La Escuelita in El Paso, Texas, especially my collaborators Lucía Durá, Victor Del Hierro, Valente Francisco Saenz, William Medina-Jerez, and Patricia Flores-Hutson, as well as our beloved families, including Brisaida Rivera, Nubia and Francisco Hernandez-Rivera, Ms. Lolita and Ximena, Heidi Rodriguez, Sergio and Brianna Flores, Ms. Cintya Gonzalez, Alejandra Dominguez, Kimberly Condia and their hermanita Valerie, Bryan Valenzuela, and Ms. Rosy. A special thank you to Judith Hernandez-Rivera for helping me coordinate all the permissions needed for chapter 4, "Language Fluidity in Health Contexts on the Mexico/US Borderland," for sharing your thoughts and feedback on this book, and for demonstrating leadership, brilliance, and kindness. Los quiero y los extraño muchísimo.

Thank you to Professor Arun Gupto for advocating for me and my work and welcoming me to the South Asian Foundation for Academic Research (SAFAR). I am also deeply grateful to the many collaborators I met through our work in Kathmandu, including Shreejana Gimire, Dipak Bastakoti, Sedu Dhakal, Karuna Karki, Shankar Paudel, Renuka

Khatiwada, Prateet Baskota, Nimesh Lansal, Nandita Banerjee, and Pragya Dahal. I am so grateful for your collaboration and friendship and look forward to much more. Thank you to Bibhushana Poudyal for introducing me to this amazing team.

I thank my colleagues at the University of Florida for always supporting me and my work and for allowing me to be in a space to write this book amid so much chaos during a pandemic. I especially thank Delia Steverson, Malini Schueller, Kenneth Kidd, Raúl Sanchez, Sid Dobrin, Marsha Bryant, Jodi Schrob, Leah Rosenberg, Rae Yan, Margaret Galvan, and Pietro Bianchi for your mentorship and support and for your warm welcome into the UF English Department family.

I am immensely indebted to Sweta Baniya, Alison Cardinal, and Emma Rose for reading and providing feedback on drafts of this book and for pushing my thinking on issues of language access and justice. Many thanks especially to Sweta and to Rukshana Kapali for your help editing the Nepali text included in this book. Thank you for being an incredible team and source of support throughout this process, in more ways than one.

What would I do without my academic family? To the ones who literally lift me up and keep me going in this game—Esther Milu, Ronisha Browdy, Suban Nur Cooley, Angela Haas, Natasha Jones, Ann Shivers-McNair, Heather Noel Turner, McKinley Green—thank you for everything, always.

I wrote a first draft of this book only weeks after I lost my mami. In times of darkness and suffering, my mami's spirit lifted me up and pushed me to do what mami always told me: to look up at the sky, remember who you are, where you come from, and why you do what you do. Thank you, mami, for always being my rock, en las buenas y en las malas, in this life and beyond.

To Victor Del Hierro, my baby, Zula, and my baby brother, Robertito, you know this, like everything, is for you and because of you.

# PREFACE

When describing community-engaged work, many researchers (myself included) have used the word *messy* as a descriptor. Community work is "messy" because people in the community do not follow the rigid, standardized research protocols we envision in the academy. Community work is "messy" because human beings inherently function outside academic boundaries designed with so few people in mind. Community work is "messy" because it requires flexibility, innovation, and humility on the part of the researcher and because it often teaches researchers that our research questions and concerns cannot and do not encompass the many malleable factors community members navigate in their everyday work. Community work is "messy" because it is real.

While the term *messy* can certainly be applicable in community-engaged praxis, this term can also be condescending and reductive when describing engagement with communities of color. As Brittany Hull, Cecilia D. Shelton, and Temptaous Mckoy (2020) explain, communities of color consistently have to function in professional spaces where their "identities, epistemologies, and even their very bodies are called into question" (7). As such, it's important for researchers to consider how labels like "messy," "loud," and "unprofessional" are grounded in white/Western/English-dominant ideals that diminish the cultural, rhetorical, and embodied practices of marginalized people.

In this book, I dig into the *messiness* of community work by unpacking what it means to do research in technical communication contexts with multilingual communities of color who communicate outside the boundaries of mainstream white English. Focusing on language relations specifically, I argue that as technical communication researchers continue to recognize the value of global research, we should grow increasingly aware of and prepared to navigate, design, and sustain justice-driven multilingual experiences. As I demonstrate through three grounded case studies in very distinct contexts, when working with communities in global contexts, technical communicators should understand not only their own positionalities but also the positionalities and histories of the languages and lands used in any research space. As academic and other institutions continue to encourage technical communication

https://doi.org/10.7330/9781646422760.c000

researchers to expand their global reach, we should consistently question what this reach entails and how our presence as English-speaking technical communicators shifts and changes the work we are able to do with multilingual communities. Focusing on language relations specifically, I argue, can provide technical communicators with an entry point into better understanding what participation can or should look like in a multilingual research space. Because language is always tied to other social factors such as race, dis/ability, gender, class, and more, grounding conversations about global technical communication in language relations can provide useful perspectives on the long-lasting impacts seemingly simple or short-term studies can have on communities.

As you read through the theoretical and methodological groundings of this book and about the brilliant communities represented in the case studies, I am sure the word *messy* will continue to emerge as a potential descriptor of the multiple disciplinary conversations being threaded together in this context. As I hope to demonstrate, to understand the impact language diversity has in technical communication research, it's important to bring together trans-disciplinary scholarship both from academia and from the multiple activist communities that have long been navigating multilingual experiences to foster access and justice. For this reason, as you move through each chapter, I encourage you not only to see the connections represented in the pages of this book themselves but also perhaps to bring further connections from your own experiences. Undoubtedly, the case studies and framing I provide in this book are hyper-localized to the contexts and communities represented. However, my goal is to push technical communication researchers to ask important questions about their own projects with multilingual communities, all with the goal of designing justice-driven methods, methodologies, and relationalities to community work that centralize language diversity as both an unequivocal reality and a powerful asset present in *all* technical communication research, whether we acknowledge it or not.

Ultimately, this book is written for technical communication researchers, teachers, students, and practitioners who work with communities to develop infrastructures, information systems, tools, technologies, and other materials to reflect community goals, values, and expertise. I wrote this book as documentation of my own journey in learning what it means to recognize, leverage, and monitor my positionality in relation to and with the people, languages, and lands with whom I have the privilege to work. As a researcher accountable to community first, I hope this book opens up and expands space for further conversation about the often ignored power dynamics embedded in all technical communication research. Furthermore, as technical communication researchers continue to practice community-engaged work, I hope this book helps us answer the question: when we say community work is *messy*, what do we mean?

# DESIGNING MULTILINGUAL EXPERIENCES IN TECHNICAL COMMUNICATION

# 1

# A "THEORY OF CHANGE" FOR GLOBAL TECHNICAL COMMUNICATION

Early Western models of communication depict the exchange of ideas and information as a simple, linear process. For example, Claude Shannon and Warren Weaver's (1949) transmission model of communication consisted of five elements: a source of information, a transmitter, a channel, a receiver, and a destination. Such linear models have been widely contested, both within and beyond the field of technical communication.

In "The Technical Communicator as Author: Meaning, Power, Authority," Jennifer Daryl Slack, David James Miller, and Jeffrey Doak (1993) argued that different models of the communication process help define the role of the technical communicator, as a transmitter, translator, or articulator (i.e., author) of meaning, and that through these models, "the place of the technical communicator—and of technical discourse itself—shifts in different relations of power" (14). In other words, the way the field of technical communication perceives the process of communication inherently influences the way the field defines the role of the technical communicator themselves—as a mere transmitter of information from one source to another, as a "translator" of information from technical and scientific audiences to "lay" audiences, or as an "articulator" who not only transmits or translates but also authors meaning. Slack, Miller, and Doak's model has also been extended and revised in technical communication research, particularly through the field's recent social justice turn, which brings more attention to the way non-Western communities have always embraced complex communication models that account for issues of power, privilege, and positionality and their role in all communication acts (Jones 2016; Jones, Moore, and Walton 2016).

While the field of technical communication continues to rightfully expand its perception of the role and power technical communicators have in facilitating communication practices and influencing the

https://doi.org/10.7330/9781646422760.c001

material realities of people and communities, much of the narrative influencing the field's definitions of communication is still rooted in a monolingual ideology, or "the notion that communication only happens through and by one language at a time" (Pérez-Quiñones and Carr Salas 2021, 66). Through this perspective, when technical communication researchers discuss the "translation" work technical communicators do, the emphasis sometimes remains on translating technical or scientific information to non-technical audiences, shifting language from one form of standardized white American English to another. Of course, many technical communication researchers do engage in conversations about translation and localization, pointing to the ways multilingual users in global contexts localize information across languages for and with their communities (Dorpenyo 2020; Sun 2012). Yet when it comes to discussing the labor of language transformation and translation specifically in general technical communication projects, some technical communication researchers revert to the (over)simplified, linear transmission models of communication the field has long contested.

For example, when we think about translation work in technical communication and related fields, we can revert to thinking of a source language uttered by an English speaker as a transmitter of information that then gets decoded by either a human interpreter or a digital translation tool before reaching its destination. While seemingly accurate, this model of multilingual communication reduces the role of the translator and of translation itself to a mere conduit of information, much in the same way early models positioned technical communicators as powerless transmitters without agency. Through this limited view of translation, one can easily ignore the experience and labor that are embedded in multilingual communication, the influence the process of translation can have on the parties tasked with this labor, and the influence of translation on the results of the communicative act as a whole. As Manuel Pérez-Quiñones and Consuelo Carr Salas (2021) clarify, "The ideology of monolingualism within the design and implementation of user interfaces not only neglects to account for the large portion of the population that is bilingual and multilingual, but by building monolingual interfaces, designers disregard the nuances of linguistic diversity and ignore the bilingual individual as a user class" (66).

Countering such oversimplification, what I present in this book is a "theory of change" (Tuck 2009) for the way the field of technical communication perceives translation and multilingual communication. When hearing the word *multilingual*, rather than thinking of a linear, automated translation process, I want to push technical communicators to recognize the entire multilingual experience, which includes not only the languages present in a communicative act but also those that are excluded and the impact these exclusions can have on all research interactions. A multilingual experience expands beyond words on paper or sentences in multiple languages, instead accounting for the emotions, histories, and embodied realities of the communicators engaged in multilingual interactions. Designing multilingual experiences, then, requires a trans-disciplinary theory of change in the way technical communicators understand the potential of language to impact their work and the responsibility we have to honor, recognize, and engage in the languages and language histories of the communities we are privileged to work with.

## A THEORY OF CHANGE

> *Theories of change are implicit in all social science research, and maybe all research. The implicit theory of change will have implications for the way in which a project unfolds, what we see as the start or end of a project, who is our audience, who is our "us," how we think things are known, and how others can or need to be convinced. A theory of change helps to operationalize the ethical stance of a project, what are considered data, what constitutes evidence, how a finding is identified, and what is made public or kept private or sacred.*
>
> —Eve Tuck, "Suspending Damage: A Letter to Communities," 413

Unangax scholar Eve Tuck (2009) urges researchers to move away from what she calls "damage-based research," or "research that operates, even benevolently, from a theory of change that establishes harm or injury in order to achieve reparation" (413). Citing studies in education that sought to increase resources for marginalized youths by documenting the "illiteracies" of Indigenous youths and youths of color, Tuck (2009) explains that damage-based research is a popular mechanism by which "pain and loss are documented in order to obtain particular political or material gains" (413). While damage-based studies have proven successful in attaining political or material gains in the form of funding, attention, and increased awareness related to the struggles of marginalized communities, Tuck (2009) points researchers to the ongoing violence damage-based research inflicts on marginalized communities, even under benevolent or perceivably beneficial circumstances. Among the many issues associated with damage-based research are the underlying assumptions this type of work makes and sustains about marginalized people; namely, that marginalized communities lack communication, civility, intellect, desires, assets, innovation, and ethics (along with much more).

To move away from damage-centered research, Tuck (2009) explains that researchers need to reorient our underlying "theories of change" regarding how we approach working within community contexts. In other words, researchers should move away from highlighting what a community is lacking or how a community has failed in the face of colonization or oppression, moving instead toward recognizing the multifaceted elements that continually (re)define a particular community as well as its knowledges and practices. As Tuck (2009) warns, if researchers only use a community's failure or oppression to justify its existence and needs for support, then these damage-based frameworks will ultimately define an entire community, consequently ignoring the multifaceted qualities all communities possess and sustain across space and time. In *Race after Technology: Abolitionist Tools for the New Jim Code*, Ruha Benjamin (2019) further explains that damage-based ideologies regarding communities of color perpetuate "coded inequity" masked as "technological benevolence" through the design of surveillance technologies that code Black, poor, immigrant, disabled communities as "unwanted," "second-class," "criminals" (9).

As a bilingual (Spanish-English) technical communication scholar who works with immigrant and transnational communities, I find the discussion of damage-based research and the push toward new theories of change relevant to contemporary research practices within and beyond the field of technical communication. In my own experiences, I have seen how multilingual communities, or communities that identify with heritage languages other than English in the US and with non-colonial languages across the world, are frequently positioned as deficient in technical and professional communication contexts. Conversations about multilingual communities are often defined around terms such as "*limited* English proficiency," directly pointing to a community's limitations in a particular language (i.e., English) without any recognition of the multiple other languages and communicative practices a community might possess.

Issues of language "proficiency" are further intensified in research related to multilingual communities of color, which are often described through their limited access to resources such as information, education, and healthcare. When talking and writing about multilingual communities of color, researchers within and beyond technical communication sometimes embrace damage-based approaches that highlight a community's struggles and even use those struggles to define that community's existence and merits. As Jeffrey T. Grabill (2007) argues, "We—as researchers, teachers, citizens—have failed to understand rhetorical work in communities as *work*," and, I would add, we consistently fail to recognize this community work as culturally responsive to long-standing histories of oppression and colonialism (2, original emphasis). This damage-based approach to multilingual research does not have to be intentional; indeed, as Tuck (2009) clarifies, damage-based approaches are often undertaken benevolently, particularly by researchers who truly do want to improve conditions and shift conversations about oppression and discrimination. The problem is that good intentions are not always paired with a recognition of the ways privilege and power shape research interactions and how these power relationships are then embedded into design. Thus, benevolent orientations to damage-based work can have dangerous consequences for communities that are "overresearched yet ironically, made invisible" (411–12). As Donnie Johnson Sackey (2020) elaborates, "There is a long history of conducting research on poor communities of color with little concern for participants' ability to control what happens with their data and whether they and their communities benefit from that data" (38). In these situations, technical communicators have an opportunity and a responsibility to move "beyond critiques of technology or user documentation in favor of designing systems that can save lives" (34).

As the field of technical communication continues to embrace its critical and groundbreaking "social justice turn"—a turn that pushes technical communication researchers to work intentionally to redress injustices and oppression (Jones, Moore, and Walton 2016; Haas and Eble 2018)—I argue that it's critical for the field to move away from damage-based orientations to multilingual research, particularly in global contexts. To do so, I suggest that we embrace, expand, and even complicate asset-based frameworks for doing community-based research in technical communication (Agboka 2013; Grabill 2007; Durá, Singhal, and Elias 2013; Haas 2012; Dorpenyo and Agboka 2018; Sun 2012; Simmons 2008; Walwema 2021) and that we apply these extended frameworks as we design multilingual experiences in, for, and with global communities (Cardinal 2019; Rose et al. 2017; Sackey 2020). As I demonstrate in this chapter, technical communicators are particularly well positioned to counter damage-based approaches to multilingual technical communication research, not simply by replacing "damages" with "assets" but rather by rejecting the notion that communication can be reduced to binaries and single-identity categories altogether. By leveraging technical communication's long-standing emphasis on making information accessible across difference, there is an opportunity for this field to further embrace complexity in working toward justice within global communication practice. As Victor Del Hierro (2019) explains, culturally and linguistically diverse communities build "complex relationships" with local technical communicators (e.g., DJs) to "create localized and accessible content" that draws on local expertise while also having global impact (28). As researchers across fields continue to acknowledge the importance and relevance of multilingualism in contemporary global contexts and as technical communicators continue to expand notions of accessibility by centering the expertise of disabled communities, I argue that technical communicators can continue to broaden what it means to create accessible communication. This type of justice-driven disciplinary expansion can only happen in collaboration with communities that have long been advocating for access and inclusion in both local and global contexts.

## EXPANDING LANGUAGE ACCESS FRAMEWORKS IN GLOBAL TECHNICAL COMMUNICATION

In their groundbreaking article "Disrupting the Past to Disrupt the Future: An Antenarrative of Technical Communication," Natasha N. Jones, Kristen R. Moore, and Rebecca Walton (2016) ask the pivotal question: "If we accept that inclusivity is an integral part of our field's history, how can or should we proceed" (212). By historicizing various movements and efforts within technical communication that push our field to further engage with and work against systems of oppression in everyday tools, technologies, and infrastructures, Jones, Moore, and Walton (2016) invite technical communication scholars to "re-envision the field" through a "larger tapestry" that not only draws on but perhaps also centralizes interdisciplinary research that expands beyond what some people may consider "traditional," "true," or "viable" technical communication work (223). To continue working toward inclusion, as many technical communication researchers have shown, it's important that we expand our disciplinary grounding, incorporating research from fields outside technical communication that have been engaged in efforts for justice and inclusion for many decades (Williams 2013). For example, Jones (2016) frames the potential of social justice in technical communication by citing scholarship across fields and disciplines, primarily by feminists of color—including work in decoloniality (hooks 1994; Smith 1999), narrative inquiry (Perkins and Blyler 1999), Black feminist thought (Collins 1999), Chicana feminism (García 1989), and Asian feminism (Bow 2011), among many others. African scholars Josephine Walwema (2021), Godwin Y. Agboka (2013), and Isidore Dorpenyo (2020) provide models for technical communicators to engage in international, multilingual research through decolonial perspectives that foster reciprocity and push toward social justice in the Global South.

In the collection *Key Theoretical Frameworks: Teaching Technical Communication in the Twenty-First Century*, Angela M. Haas and Michelle F. Eble's (2018) contributors further illustrate the importance of centralizing interdisciplinary research when imagining possibilities for social justice in technical communication, highlighting research on feminism (Frost 2018), Indigenous rhetorics (Agboka 2018), hip hop (Del Hierro 2018), and queer theory (Cox 2018) to offer new possibilities for the ways technical communication as a field can continue to expand its boundaries in its efforts toward justice and inclusion. As Haas (2012) explains in her discussion of how she developed a course on race, rhetoric, and technology in 2009, at a time when the field of technical communication was not yet as "enriched by the recent conversations about race and ethnicity emerging in the discipline today," she opted to "piece together a curriculum I could believe in by weaving together the scant threads of inquiry on race, rhetoric, and technology in our field with some of the existing strands in cultural, critical race, rhetorical, and feminist studies" (278). This inter- and trans-disciplinary framing within the context of technical communication pedagogy, Haas (2012) elaborates, "provided me and my students with multiple places to stand in the field at connected but different intellectual intersections" (278). These intersections, I argue, are critical to the growth and sustainability of social justice–driven initiatives within and beyond technical communication, particularly as our field continues to work with and in global contexts.

As a language and translation scholar in the field of technical communication, I also blend and expand disciplinary groundings to establish "multiple places to stand in the field" (Haas 2012, 278). As technical communication scholars continue to engage in global research with marginalized communities, I suggest that we expand the field's model of communication to re-imagine what it means to provide language access, leveraging interdisciplinary conversations across language and disability studies to envision new futures for designing accessible content alongside communities.

## (RE)SITUATING LANGUAGE DIVERSITY AND MULTILINGUALISM IN TECHNICAL COMMUNICATION

Technical communication researchers understand that tools, technologies, and systems can no longer be designed in standardized white English alone, and we are working to recognize that designing in multiple languages should be a common practice that takes place throughout, rather than after, our initial design and prototyping processes (Batova 2018; Cardinal 2019; Rose et al. 2017). In addition, technical communicators are increasingly acknowledging the value and importance of working with translators and interpreters to make information accessible not only to English speakers (Batova 2010, 2018; Walton, Zraly, and Mugengana 2015), and we largely understand the value (both materially and ideologically) that comes with our efforts to engage in multilingual communication. Our field has also begun paying more attention to the way technical communication work happens with Indigenous, transnational, and immigrant communities and other historically marginalized groups with expertise in multilingual communication within and beyond the US. In this work, some researchers emphasize the important role of social justice in working with linguistically and ethnically diverse communities, highlighting the ways globalization efforts can render colonizing violence (Agboka 2013; Haas 2012) when human dignity and human rights (Walton 2016) are ignored (Jones and Williams 2018; Williams and Pimentel 2014). In short, as a profession that has historically been described through translation metaphors and as the experts and partners in engineering who "translate" techno-scientific information for lay audiences (Slack, Miller, and Doak 1993), technical communicators are well poised to recognize and value the importance of language diversity in relation to information access and technical content creation, particularly in the context of globalization (Batova 2018; Haas and Eble 2018; Walwema 2016, 2021).

Working from an understanding that technical communication as a field *already* recognizes the value and importance of globalization (in terms of both language and design), my goal in this book is to help our field further connect issues of globalization and language access to the bodies, communities, and lands through which globalization, translation, and internationalization happen (Agboka 2013, 2018; Durá, Singhal, and Elias 2013; Haas 2012; Dorpenyo and Agboka 2018; Sun 2012; Walwema 2021). As Tatiana Batova (2018) explains, in current global contexts, "while technical translation is included under the umbrella definition of TC [technical communication], relations between the professionals in these two fields are most often those of contractors and clients, and the communication between these groups is far from perfect" (79). In other words, while technical communication as a field acknowledges the value of globalization and the importance of translation in fostering global reach, the roles, expertise, and experiences of translators could be further highlighted within technical communication scholarship and practice. As Haas and Eble (2018) further clarify, "Globalization—and the complex and culturally-rich material and information flows that come with it—has forever changed who we think of as technical communicators, the work that technical communicators do, and where and how we understand technical communication happens" (3). My goal in this book, then, is to also illustrate not only how globalization has changed us as technical communicators but also how we as technical communicators, through language-driven relations specifically, change the communities we inhabit and could further support the good work and technological change that is already taking place in the world (Durá, Singhal, and Elias 2013; Grabill and Simmons 1998; Grabill 2007; Shivers-McNair and San Diego 2017; Walton and Hopton 2018). As Haas and Eble (2018) continue, "While technical communicators may appreciate the international, professional, and economic gains afforded to us by globalization, we must also interrogate how we may be complicit in, implicated by, and/or transgress the oppressive colonial and capitalistic influences and effects of globalization" (4).

In working toward "transgress[ing] the oppressive colonial and capitalistic influences and effects of globalization" (Haas and Eble 2018, 4), I argue that technical communication researchers should acknowledge the intersectional identities of global communities in situated contexts and expand our notions of language access accordingly. As Allison Hitt (2018) argues, technical and professional communicators should develop programs, spaces, and pedagogies "that acknowledge the rhetorical situatedness of accessibility" and that recognize how centering disability in design can benefit both disabled and non-disabled audiences (53). While the field of technical communication has long understood the complexities and necessities of designing information in languages other than English for and with international and transnational audiences (Batova 2010; Maylath and St.Amant 2019; Rose et al. 2017; Williams and Pimentel 2014), we have a lot of work to do in expanding how we understand the intersecting identities of these audiences and how we recognize the often invisible and embodied labor encompassed within a seemingly simple concept like translation. For this reason, as technical communication researchers continue to innovate new methodologies for navigating the "messy," "complex" nature of conducting technical communication work in multilingual contexts (Walton, Zraly, and Mugengana 2015) and as our field continues to pay more attention to the development of technical tools and documents that are accessible in languages other than English (Rose et al. 2017; Walton and Hopton 2018; Walwema 2021), I argue that we should continue to expand our understandings of language access to further centralize the bodyminds (Price 2011; Schalk 2018) of multilingual communities and that we should recognize the identities and methods of participation of global audiences beyond single-identity categories and binaries. To this end, this book positions language diversity and translation specifically as critical components of technical communication, urging researchers in the field to recognize the embodied nature of language and the complex process of language transformation as part of a broader multilingual experience.

When understood through what I call an *intersectional and interdependent* orientation, a methodology that draws from both critical race studies and disability studies, language diversity can be used as a point of analysis, intervention, and collaboration in technical communication—allowing researchers to continue to develop critical frameworks for designing and sharing tools, technologies, platforms, pedagogies, and projects that work toward language access. At the same time, I argue that accomplishing language access is just one component of designing a successful multilingual experience and that technical communication researchers should continue to acknowledge their own positionalities when working with communities that do not communicate predominantly in standardized white English (Jones and Williams 2018). By pairing an emphasis on translation with a broader understanding of multilingualism that is contextualized through interdisciplinary frameworks, the central aim of this book is to help our field reframe and rethink the methodologies, practices, and ideological commitments we often associate with language diversity.

## NEW APPROACHES TO ESTABLISHED FRAMEWORKS IN MULTILINGUAL TECHNICAL COMMUNICATION

In her foundational work *Cross-Cultural Technology Design*, Huatong Sun (2012) explains that when cross-cultural work is approached through positivist orientations that value efficiency over depth, researchers can (sometimes unintentionally) address only the "tip of the iceberg" in cross-cultural interactions. That is, when cross-cultural design research relies on practical checklists and standardized protocols, researchers can adopt or even reproduce cultural stereotypes (Ding 2020; Jones and Williams 2018) instead of localizing technologies and designs successfully across cultures and contexts. Similar arguments have been made by disability studies scholars who argue that access in technical communication design should not be reduced to simple checklists and protocols, since notions of access should be rhetorically situated and localized (Hitt 2018; Yergeau et al. 2013; Zdenek 2015). Sun (2012) proposes "user-localization" as an approach to cross-cultural technology design that values local community knowledge and emphasizes the role of local expertise in localization practices. Other researchers, such as Huiling Ding (2020), explain that "participatory, user-centered design can play important roles in user-empowerment and ethical engagement with users in civic, educational, and industrial settings" (145). Walwema (2021) further emphasizes the importance of paying attention to not only how information is designed but also how it is distributed in transnational contexts through the use of social media strategies that serve "the rhetorical circulatory function of rallying the public" (130). While this research focuses on the importance of intercultural communication and accessibility in technology design and dissemination, I argue that in many technical communication projects, translation and multilingualism are sometimes treated through similar reductive, positivist orientations that dismiss the role race and embodied difference play in multilingual technical communication practice (Gonzales 2018; Batova 2010).

Let me give you some examples.

When engaging in technical communication research, many US-based technical communicators are already in the practice of collaborating with language interpreters, or individuals who translate verbal information across languages. US-based technical communication researchers may employ professional interpreters when conducting research abroad, relying on these professionals to translate information among researchers, participants, and other stakeholders. As some scholars have noted, working with language interpreters adds a layer of richness and complexity to technical communication research while also allowing US-based researchers to navigate intercultural issues when working with transnational or international communities (Batova 2010; Hopton and Walton 2019; Walton 2016; Walton and Hopton 2018; Rose et al. 2017). In some cases, research teams already include bilingual or multilingual members who can facilitate communication between participants and researchers who do not speak the same language(s), providing additional insight and expertise in projects that take place in multilingual contexts. In other cases, as has been the case with my own research with Spanish-speaking communities (Gonzales 2018), bilingual or multilingual technical communication researchers engage in work with their own heritage language communities, translating interactions for English-based audiences in presentations and publications while conducting most of our work in languages other than English.

Perhaps less ideally, technical communication researchers sometimes rely on English-speaking members of their participants' communities to translate information for the duration of a study, even when the participants do not have training in professional interpretation and, in some cases, when community members are not paid for their interpretation work. As part of the realities and contexts in which contemporary multilingual technical communication work currently happens, technical communicators may give the responsibility of translation to a participant, a friend, or, frankly, anyone who can help facilitate communication so the research can continue and the research team can maintain focus on the "real" purpose of the project (e.g., conducting an interview or focus group). While I am not saying that these practices are always unethical or unsuccessful, I do argue that in any of the aforementioned frameworks, when language diversity is viewed as a methodological issue or problem to be solved rather than being conceptualized as a central component of the interactions that frame an entire project or study, technical communication researchers may (perhaps unintentionally) miss important perspectives or elements involved in a project that can significantly impact both the results of a research project and the impact our field has on already marginalized and misrepresented or unrepresented communities. As disability studies scholars have long advocated, ignoring the invisible, embodied elements embedded in all research methodologies can erase rather than highlight important experiences in research (Price and Kerschbaum 2016).

In many ways, the general concepts, methodologies, and ideas presented in this text are nothing new to technical communication, as they can all easily fall under areas of study already common in technical communication research—including, for example, participatory design, human-centered design, international or intercultural technical communication, action-based research, user experience, civic engagement, service learning, and perhaps even less apparent areas such as risk communication, medical rhetorics, and disability studies. However, through grounded case studies of multilingual technical communication, I argue that when we consider and engage with the presence of language diversity in all these areas of study through our work as technical communicators, we can better account for how our work is positioning and being positioned by our various stakeholders and participants.

This book illustrates how a focus on language in technical communication research can help researchers in the field better understand our work and its impacts on the people we seek to communicate with and through. In the cases of technical communication researchers who work in multilingual communities in collaboration with language interpreters, the case studies presented in the following chapters can help researchers answer general questions such as:

- (How) does the fact that my communication with research participants is taking place through an interpreter influence the potential findings of this study?
- (How) does or can my presence as a researcher who does not speak the same languages as my participants influence my participants' comfort with and trust in our interactions?
- (How) can I collaborate more successfully with both my multilingual participants and my interpreter(s) so I contribute more directly to the language labor of this project?

Furthermore, for researchers interested in designing multilingual tools and technologies alongside multilingual communities, the case studies presented in this book can provide some possibilities for questions such as:

- How can tools and technologies be designed as inherently multilingual rather than being retrofitted for language access purposes?
- How can researchers engage in collaborative design activities to generate ideas with multilingual communities in culturally supportive and sustaining ways?
- What does it mean to conduct research in English with communities for which English is tied to long-standing histories of oppression and colonialism?

In addition to providing some possible answers to these questions, the case studies presented in this book also collectively illustrate the fact that interpretation (i.e., the verbal transformation of information across languages) and translation (i.e., the written transformation of information across languages) are perhaps the most visible and recognized aspects of multilingual technical communication experiences, but they are not the only elements involved in successful language access. That is, while much research has highlighted the importance of translation and interpretation in technical communication work (Walton, Zraly, and Mugengana 2015; Maylath and St.Amant 2019), I argue that more attention should be paid to other elements of multilingual technical communication experiences, including the connections between language and race, the influence of language on researcher and participant positionality, and the connections among language, power, land, and materiality in collaborative technical communication research. In the sections that follow, I provide an outline of the chapters in this book while also highlighting the interdisciplinary theoretical and methodological frameworks that influence these projects.

## STRUCTURE OF THIS BOOK

To define what I mean by "multilingual experiences in technical communication," I begin by threading together theoretical and methodological frameworks within and beyond the field of technical communication. To this end, in chapter 2, "An Intersectional, Interdependent Approach to Language Accessibility in Technicial Communication," I present what I call an intersectional, interdependent approach to accessibility in technical communication, where I bring together scholarship in racial and linguistic diversity with scholarship in disability studies to argue that when working in multilingual environments, technical communication researchers should consider language and translation through an intersectional perspective that considers access beyond their own positionalities as researchers. As social justice and disability studies scholars have taught us, accessibility in any environment is a shared responsibility. As such, relying on translators, interpreters, participants, and community members to "handle" or "take care of" all language-access work renders an oppressive power dynamic among researchers, communities, and language professionals. As I demonstrate in chapter 2, orienting to multilingual technical communication work through intersectional, interdependent frameworks can allow technical communication researchers to gain important insights into their work while also taking more responsibility for the impact our work has on our communities and surrounding lands and environments (Agboka 2018; Sackey 2020).

Based on the introduction of this intersectional, interdependent framework, chapter 3, "Research Design," sets up the research design for this project, where I outline the specific methods I co-selected with my collaborators and participants to study what it means to conduct technical communication research in multilingual contexts. In this chapter, I connect my project to ongoing work in technical communication in areas such as human-centered design, user experience, participatory design, and localization. I also introduce the research questions this book seeks to answer, which predominantly consist of: what does technical communication look like in multilingual contexts, and how can technical communicators design multilingual experiences that benefit (rather than ignore or harm) the multifaceted identities and experiences of linguistically and ethnically diverse communities? By asking these questions, I argue that technical communication researchers can continue working to design multilingual technical communication experiences in our research and professional spaces as well as in our classrooms. In this chapter, I also begin to introduce readers to my various research participants and collaborators, which include health-related organizations in the borderland city of El Paso, Texas, a research center and community organization in Kathmandu, Nepal, and a legal services and activism organization developed for and by Indigenous language interpreters and translators in Oaxaca City, Oaxaca, Mexico.

Professional translators and interpreters often receive training in one or more of the following areas: (1) medical translation and interpretation, which can encompass the written translation of medical documents such as patient medical history forms, medical terminology documents, and health campaign documents and/or the verbal interpretation between healthcare providers (e.g., doctors, nurses, medical personnel) and patients who speak various non-dominant languages; (2) legal translation and interpretation, which can encompass the written translation of legal documents such as court proceedings and decrees and/or the verbal interpretation between legal staff (e.g., lawyers, judges) and members of the public who speak various non-dominant languages; and (3) community translation and interpretation, which can encompass the written translation of any document used in community interactions (e.g., flyers for and information on community events, business plans, annual reports) and the verbal interpretation between community members and various business personnel or organization employees who serve the public (e.g., local library staff, social workers, teachers). These three areas are identified by professional translation and interpretation organizations as the places where linguistic movements mitigate human activity, and, as such, the case studies I present in this book are structured around these same areas. By presenting case studies in medical, legal, and community contexts, I seek to illustrate how technical communicators and professional interpreters and translators share common ground, interests, and responsibilities in the creation and sustainability of multilingual experiences for a wide range of stakeholders.

I begin my case studies with chapter 4, "Language Fluidity in Health Contexts on the Mexico/US Borderland." In this chapter, I illustrate how language ideologies shape the way people experience health and healthcare, particularly in a borderland community that moves fluidly across many variations of Spanish and English in everyday interactions across contexts. In this community, providing translations of medical information in standardized English or standardized Spanish is an ineffective strategy that will not reach community members who prefer using both Spanishes and Englishes to communicate in verbal and written forms. In collaboration with organizations in El Paso, Texas, that seek to provide access to healthcare for binational and bilingual community members in both El Paso and the neighboring city of Ciudad Juárez, Chihuahua, Mexico, this chapter provides insights into the complexities of designing multilingual experiences around technical medical documents and information that does not easily fall into a single linguistic category. The chapter introduces scenarios, conflicts, and possible strategies for designing language access experiences around issues of health and wellness in contemporary contexts where language categories are fluid and constantly evolving.

Moving from the borderland community of El Paso and Ciudad Juárez to a research organization in South Asia, chapter 5, "User Experience and Participatory Design in Kathmandu," presents findings and narratives from a participatory design workshop series I co-facilitated at the South Asian Foundation for Academic Research (SAFAR), an independent research center located in the city of Kathmandu. In collaboration with an interdisciplinary team of researchers and students in both the US and Nepal, this project sought to illustrate what participatory design can entail in a South Asian context, particularly with students and professionals committed to shifting representations of their own languages and cultures in online spaces. By introducing an ongoing collaboration with this research center and its various academic, industry, and community stakeholders, this chapter illustrates how common participatory design and user experience methods and methodologies (e.g., usability testing, journey mapping, affinity diagramming, prototyping) can be adapted in communities that span widely across linguistic, cultural, and national boundaries. Multilingual experiences in participatory design projects, as this chapter demonstrates, should be grounded in community expertise and community values to render results that are localized and effective rather than merely performative.

Extending from a community-based participatory design project to multilingual experiences in legal realms, chapter 6, "Linguistic and Legal Advocacy with and for Indigenous Language Interpreters in Oaxaca," details a collaboration with the Centro Profesional Indígena de Asesoría, Defensa, y Traducción (CEPIADET), an organization developed, led, and sustained by Indigenous language interpreters who specifically advocate for the representation of Indigenous languages and Indigenous language interpreters in legal processes within and beyond Mexico. By introducing ongoing work with this organization, including the design and development of an international gathering that brought together 370 Indigenous language interpreters and translators from Mexico, Peru, and the US in Oaxaca, this chapter illustrates ongoing efforts by interpreters of Indigenous languages who work in legal settings and who navigate legal communication in Spanish and various Indigenous languages—including variants of Mixe, Mixteco, Zapoteco, Nahuatl, and Quechua. Through a discussion of interviews, community events, and collaboratively designed documents stemming from an organized gathering of legal interpreters, this chapter provides strategies for designing multilingual technical communication experiences in languages that stem beyond Western notions of communication and that encompass continual relationships with community members as well as their surrounding lands and environments. In this chapter, I argue that multilingual technical communication experiences, when designed through collaborative, justice-driven models, can have impacts not only on individual people but also on the preservation of intergenerational knowledge and environmental sustainability.

The ultimate goal of putting these projects together is to illustrate that multilingual experiences in technical communication can span languages, cultures, contexts, and communities while still being grounded in participatory methodologies that center the expertise of linguistically and ethnically diverse communicators. By focusing on linguistic movements (i.e., translation, interpretation) in the analysis of disparate projects across different areas, including medical, legal, and community contexts, I illustrate how multilingual experiences are critical to the work of contemporary technical communicators across various areas of specialization. In chapter 7, "Implications for Designing Multilingual Experiences in Technical Communication," I draw from the case study data presented in previous chapters to share practical strategies technical communication researchers can enact in their work with linguistically and ethnically diverse communities. Specifically, I argue that centralizing multilingualism as a critical component of contemporary technical communication can help researchers in the field continue to recognize how language shapes and influences the impact our work can and should have in particular communities and environments. In this chapter, I conclude by arguing that technical communicators should continue to recognize language diversity not only as an issue that needs to be navigated to do the work of technical communication but also as an asset that can help technical communicators expand the ways through which we can mitigate communication and make information accessible across cultural, linguistic, and national borders. While the work of technical communication has always encompassed language diversity, paying closer attention to how our language is shaped and transformed across linguistic, racial, and cultural contexts can help technical communicators continue to work toward justice and equity in our research, pedagogical, and professional practices.

Attuning to language difference through justice-driven frameworks requires added attention to relationality (Collins 2019)—between languages and cultures, bodies and spaces, communities and practices, dis/abilities and design. In chapter 2, I illustrate how technical communicators can embrace the possibilities of language diversity and its role in our field through multiple perspectives that expand a single axis of identity. To do so, I suggest that technical communicators thread together methodological practices from both critical race studies and critical disability studies to embrace a methodology for designing multilingual experiences that center difference as the core of successful and accessible design. Through an intersectional, interdependent methodology, multilingual technical communication positions difference not as a problem to solve or as a possibility for tapping into new markets but as an opportunity to seek new collaborations, understandings, and innovations by designing with and for culturally and linguistically diverse communities.

As I further demonstrate in chapter 2, intersectionality (Combahee River Collective 1977; Crenshaw 1989) allows researchers to centralize race while accounting for the intertwining layers of experience, history, power, and positionality that take place as individuals navigate communication across communities and languages. Interdependency, through an emphasis on access and inclusion, provides "an ethic for intellectual work" in which participants, researchers, and other stakeholders involved in a project can take an active role in making communication accessible for all those involved (Jung 2014, 101). Thus, orienting to language diversity through intersectional, interdependent approaches can provide an avenue to move away from damage-based perspectives that position language difference as deficit. Intersectional, interdependent methodologies can also, and perhaps more importantly, provide technical communicators and other researchers with a methodology for listening to the people, bodies, communities, and lands that make multilingual communication possible and accessible across contexts. As you read about an intersectional, interdependent approach to multilingual technical communication, I encourage researchers and practitioners to think about the multiple intersecting identities of technical communication audiences in global contexts, considering how global audiences and communities can more directly shape future directions in our collective work.

# 2

# AN INTERSECTIONAL, INTERDEPENDENT APPROACH TO LANGUAGE ACCESSIBILITY IN TECHNICAL COMMUNICATION

*If I'm having a pain day and a hard time processing language and I need you to use accessible language, with shorter words and easiness about repeating if I don't follow, and you do, that's love. And that's solidarity. If I'm not a wheelchair user and I make sure I work with the non-disabled bottom-liner for the workshop to ensure that the pathways through the chairs are at least three feet wide, that is love and solidarity. This is how we build past and away from bitterness and disappointment at movements that have not cared about or valued us.*

—Leah Lakshmi Piepzna-Samarasinha, *Care Work: Dreaming Disability Justice*, 75

Simplifying complex information is at the core of what technical communicators do. Once understood as the "companion[s] to engineering" (Hart-Davidson 2001, 145) who translate information for lay audiences (Slack, Miller, and Doak 1993), technical communicators' history is in many ways rooted in an efficiency model that values clarity and simplification in communicative practice (Jones 2016). Yet for almost as long as the field of technical communication has been established, researchers and practitioners have been calling attention to the complexity of the profession, and many are more recently recognizing that making information accessible to broad audiences requires an attunement to issues of both ethics and justice. As Natasha N. Jones and Miriam F. Williams (2018) argue, "Technical communication scholarship, when wrestling with communication that fails the audience, has been mostly concerned with well-meaning and unintentional mistakes in technology, text, graphics, and document design. While the general perception that communicators and designers have their users' best interests at heart can be a positive and productive starting point, an acknowledgment of more sinister and cynical purposes for communication design

https://doi.org/10.7330/9781646422760.c002

is also necessary" (372). As I illustrate in this chapter, the simplification of information has long been centralized in technical communication and the field's emphasis on accessibility. Yet as the field continues to seek to move toward justice-oriented frameworks that value and sustain diversity, researchers should continue to develop methodologies for embracing complexity, rather than aiming for simplification, to avoid harm in the pursuit of effective communication. Embracing complexity is central to any "movement-building framework" that works toward justice by centering the strengths, needs, and desires of marginalized communities (Piepzna-Samarasinha 2018, 15).

In her introduction to *Communicating Race, Ethnicity, and Identity in Technical Communication*, Williams (Williams and Pimentel 2014) explains that while the field of technical communication continues to expand, particularly alongside the development of emerging information technologies, "still, and unfortunately, we lag behind our colleagues in other areas of English studies (literature, rhetoric and composition, and creative writing) in finding ways to wrestle with two core elements of American identity—race and ethnicity" (1). Indeed, as Williams (Williams and Pimentel 2014) further elaborates, race and ethnicity are "elements of our identity [that] shape user experiences as much as education, literacy, gender, nationality, or any of the other criteria we use to analyze audiences" (1). Jones (2016) continues emphasizing the need for technical communicators to continue to address issues of race and ethnicity in their work, especially in relation to issues of social justice. As Jones (2016) argues, "Scholars in technical and professional communication (TPC) are beginning to recognize contemporary exigencies and acknowledge that they cannot sit idly by in the midst of such sociopolitical and socioeconomic strain" (343). Engaging with issues of race, ethnicity, and justice requires that technical communicators move beyond an efficiency model that emphasizes simplicity to embrace the *complexity* and nuances embedded in and sustained by diverse audiences, communities, clients, and environments. While technical communicators continue to work toward information access, as we have always done, we are finally grappling with the fact that access means different things to different bodies across different contexts and in different lands. What the field of technical communication continues to need, then, are more models for engaging with communicative complexity through frameworks that centralize and expand conceptions of access and social justice (Haas and Eble 2018; Walton, Moore, and Jones 2019).

Language diversity and translation are elements of contemporary technical communication practice that if implemented and embraced through social justice–driven orientations can help the field work toward accessible communication that is not only clear, concise, and accessible to broad audiences but is also respectful and representative of these audiences' socio-cultural backgrounds, dis/abilities, and rhetorical goals. As I will illustrate through the case studies presented in this book, multilingualism is an avenue through which technical communicators can interrogate their own positionalities and their relationships to diverse audiences, communities, and publics (Jones and Williams 2018). The key to doing this successfully, in my perspective, is orienting to technical communication work through a critical awareness of the role the technical communicator plays in establishing and shifting power dynamics with various stakeholders. As technical communicators, we cannot assume that our presence or involvement in technical communication work will be neutral or inconsequential (Jones 2016). This includes, for example, assuming that the translation of technical documents will not influence the content of those documents and/or that the presence of an interpreter in a multilingual project will not impact the dynamic of our research and the impact of this work on specific communities and their relationships (Walton, Zraly, and Mugengana 2015).

Instead, I argue that as technical communicators, we should recognize how our involvement in contemporary technical communication practice—whether it takes place in a community, behind a screen, or both—will undoubtedly impact what we design and distribute, and we should work continuously to ensure that our impact avoids harm as much as possible. In my experience, particularly in designing multilingual technical communication across contexts, this type of orientation also forces technical communicators to recognize when we do cause harm, acknowledge these mistakes, and continue working to shift our methodological approaches toward justice. This shift from understanding our positionalities and their influence on power dynamics in our research to then recognizing our failures and working toward redressing harm is, for me, the type of intentional move that extends critique and analysis to action—what Jones, Kristen Moore, and Rebecca Walton (2016) identify as a key component of social justice praxis.

In this chapter, I introduce a methodological orientation to engaging in multilingual technical communication through what I call an intersectional, interdependent methodology. Drawing on critical race studies, specifically in connection to raciolinguistics, in combination with disability studies' perspectives on accessibility and embodiment, I provide strategies for technical communicators who want to engage in multilingual technical communication by expanding conceptions of language access through intersectional perspectives. To frame this discussion, I want to start by recognizing that, as Margaret Price and Stephanie L. Kerschbaum (2016) note, no method or methodology is fully accessible or ethical; and as researchers, we are always making choices that both foster accessibility for some audiences and limit accessibility and representation for others. Thus, when I present an intersectional, interdependent methodology, I am in no way arguing that the work I share in this book is fully intersectional and fully accessible through interdependent approaches. Instead, as I hope to demonstrate both in this chapter and in the case studies I share in chapters 4–6, an intersectional, interdependent methodology allows me as a researcher to recognize the broad landscape of languages, races, cultures, and dis/abilities present in my community work and to then negotiate the ways my own positionality and background will undoubtedly influence the spaces I inhabit as both a researcher and a community member in relation to other people, communities, and lands. Thus, to introduce the broad landscapes of communication that should be recognized in multilingual technical communication research, I'll begin by sharing a story of a recent community project in which I saw how issues of racial and linguistic difference intersected with participants' dis/abilities. Rather than positioning these intersections of difference as "barriers" or "deficits" or even claiming to define differences as "assets" to communication design and multilingual technical communication research, I use this introductory story merely to illustrate the interlocking (Crenshaw 1989) layers of difference that are always present, yet often overlooked, in contemporary research contexts.

## A DAY AT THE BRONCO

On a Saturday afternoon in October 2018, while I was living in El Paso, Texas, I (along with a collaborator) drove to a local flea market in town to gather survey responses for a project related to health disparities on the Mexico/US border. While I will introduce this project in more detail in chapter 4, what is relevant here is the fact that I went to the flea market with the goal of collecting twenty survey responses from local community members. This flea market, called the "Bronco," was located in the lower Valley region of El Paso, which is characterized as encompassing primarily working-class first-, second-, and beyond generation Mexican Americans. Given the large Spanish-speaking population in El Paso in general and in this part of the city specifically, I brought two sets of surveys with me—one in Spanish and one in English. "Simple,"

I thought. I would just ask participants in what language they preferred to read the survey, and then I would be prepared to give them a survey in their preferred language.

As we walked around the flea market looking for people who would help with this project by taking a survey in exchange for a Walmart gift card, we encountered a small booth with trinkets for sale, where several men sat around talking as they encouraged customers walking by to stop by and check out the merchandise. As we walked by the booth, I stopped and asked the men in my best researcher voice:

> "Buenas tardes. Estamos conduciendo una encuesta sobre la salud aquí en El Paso. ¿Sera que me pueden ayudar con este estudio? Solo tendrían que completar una encuesta que se demora aproximadamente quince minutos. Les puedo ofrecer un reconocimiento en la forma de una tarjeta de regalo de Wal-Mart por su ayuda" (*Good afternoon. We're here conducting a health-related survey about the community here in El Paso. Would you be willing to help me with this study? You would just have to complete a survey that will take you about fifteen minutes to finish. I can offer you a gift card to Wal-Mart in exchange for your help*).
>
> Eduardo (pseudonym), a blind man in his early sixties, responded for the group and said, "Yes, mija. We can help you out. But can you read it to me? Esque no puedo ver (*I can't see*)."
>
> Eduardo pointed to his eyes and smiled.
>
> "Claro que si," I said, "Do you want me to read it in English or in Spanish?"
>
> "No, no, English is good," replied Eduardo.

I proceeded to read the survey aloud to Eduardo while my research collaborator passed out surveys to the other men at the same booth. Although all the men in this booth spoke in Spanglish (i.e., a fluid combination of Spanish and English) to me and in Spanish to each other, they all wanted to take the survey in English. As I began reading to Eduardo, I noticed that the other men put down their pencils and began listening, asking me for clarification as I read each question-and-answer option. Rather than taking the survey individually, this group of men eventually took the survey together, walking through each question slowly as I read aloud. While in this specific instance Eduardo was the only participant who voiced wanting the survey to be read aloud, every person in this group benefited from the verbal instructions and paused me along the way to translate specific words into Spanish when the English-language descriptions of terms used on the survey were not easily understood. As the men listened to the survey, they asked each other questions, made clarifications, and, in the end, gave me several suggestions regarding which terms could be better explained on the survey itself. I left the flea market that day with all my surveys completed but, more important, with a new perspective on this survey research method as a whole. I left asking myself so many questions, including:

- Why did I assume that all my participants would be able or even want to read the survey?
- Why did I separate English and Spanish on two separate surveys rather than use Spanglish?
- Why did I approach Eduardo and the group of men in Eduardo's booth by speaking in Spanish?
- Why did I assume that these men would prefer to speak in Spanish rather than English?

The answers to these questions are as simple as they are difficult to admit. I made these assumptions about my participants because of my own raciolinguistic ideologies and my own assumption about the "default users" who would engage with my survey. Despite my very best intentions and attempts at being ethical, I projected oppressive ideologies onto this project—ideologies my participants helped me, through their labor and time, better understand. Let me unpack this a bit further.

In "Unsettling Race and Language: Toward a Raciolinguistic Perspective," Jonathan Rosa and Nelson Flores (2017) "interrogate the historical and contemporary co-naturalization of language and race" to describe what they term "raciolinguistic ideologies" (622). Through this discussion, Rosa and Flores (2017) tie the separation of communicative practices into categorical "named" languages (e.g., Spanish, English, French) directly to a broader colonial project. Drawing on Sinfree Makoni and Alastair Pennycook (2006), Rosa and Flores (2017) explain that "in conjunction with the production of race, nation-state/colonial governmentality imposed ideologies of separate and bounded languages on colonized populations" (623). In other words, European colonization established binary categorizations among countries, nations, and languages—all as part of a broader project intended to maintain white supremacy. Colonization (i.e., colonizers) enslaved Black and Indigenous people and separated lands into nations, people into racial categories (where white European is superior and all Others are inferior), and languages into static, bounded practices that were either literate/legible or not—all based on a white European standard. For example, "European colonizers described indigenous language practices as animal-like forms of 'simple communication' that were incapable of expressing the complex worldviews represented by European languages" (624). This distinction between "simple" or "animal-like" communication and the "complex" or "sophisticated" language of the European colonizers continued to fuel the dehumanization of racialized subjects through chattel slavery

(Makoni and Pennycook 2006) and has an extended and deep-rooted influence on what is deemed "complex" or "sophisticated" versus "lay" or "plain" language today (Jones and Williams 2017, 2018). In other words, as Rosa and Flores (2017) continue, "colonial distinctions within and between nation-state borders continue to shape contemporary linguistic and racial formations" as well as the methodologies and ideologies through which many contemporary Western researchers study, design, and practice communication (623).

Raciolinguistic ideologies, as described by Rosa and Flores (2017), Flores and Rosa (2015), Sami H. Alim, John R. Rickford, and Arnetha F. Ball (2016), and Esther Milu (2021) essentialize racial and ethnic identity. Through the perpetuated and often deemed accurate by default assumptions of the "white listening subject" (Rosa and Flores 2017; Flores and Rosa 2015), specific languages are by default connected to specific enthnoracial groups (Milu 2021) who are presumed to have specific levels of language proficiency in their presumed preferred languages. As April Baker-Bell (2020) explains, "Sociolinguists and language scholars have for decades described the harm an uncritical language education has on Black students' racial and linguistic identities," particularly when Black students are denied "the right to use their native language as a linguistic resource during their language and literacy learning" (8). These uncritical approaches to language education foster long-standing prejudices against Black, Brown, and Indigenous people who will never speak "good" English despite any level of education. At the same time, white subjects can embrace and use the linguistic expressions of Black, Brown, and Indigenous communities freely and fully without being regarded as less "literate," "clear," or "sophisticated." Furthermore, raciolinguistic ideologies and uncritical approaches to language education contribute to the ongoing erasure of Indigenous languages in preference for Western, white language use across the world (Cusicanqui 2010).

So, what does this have to do with my assumptions about survey participants at the flea market? When I approached a group of Brown Mexican men at a flea market in one of the poorest areas of the borderland city of El Paso, Texas, I spoke in Spanish because of my own raciolinguistic ideologies that connected my participants' embodied positionalities to a language practice (i.e., Spanish) that I (a white listening subject on this occasion) assumed would be most appropriate for this particular rhetorical context. My own language practices and presumptions led me to believe that I would be facilitating conversation with and perhaps even honoring the language practices of my participants by using my knowledge of Spanish to begin this conversation. Yet, as Eduardo and the other men in his flea market booth that day showed me, I made these assumptions without recognizing the broad linguistic practices of the borderland region and without acknowledging the racist, colonial ideologies that have worked for centuries to actively erase Spanish and multiple Indigenous languages from being spoken on that colonized land. Raciolinguistic ideologies may connect Brown Mexican bodies in the US to Spanish, erasing the impacts of racialization, migration, colonization, ongoing racism, and general language plurality that exist in the world today.

In addition to making assumptions about Eduardo's language preferences, I also made assumptions about how my participants, both in Eduardo's booth and in general, would access my survey questions. Many researchers and research contexts default to ableist assumptions about who is accessing research protocols and how users engage with methods such as interviews, surveys, and ethnographies (Price and Kerschbaum 2016), as well as with public spaces in which research is shared (Piepzna-Samarasinha 2018). In their discussion of "interviewing sideways, crooked, and crip," Price and Kerschbaum (2016) "challenge the assumption that a semi-structured interview should proceed like an oral/aural conversation, and that the ideal interviewer should be conventionally abled—or even super abled" (22). Instead, as many disability studies scholars both within and outside of technical communication recognize, research protocols and methods should be designed with dis/abled communities in mind, both because this is ethical, justice-driven praxis and because, as Cynthia L. Selfe and Franny Howes clarify (in Yergeau et al. 2013), "what's good for people with disabilities often ends up being good for everyone" (n.p.).

In my interactions with Eduardo, the reading aloud of questions brought the men in Eduardo's booth together to take my survey collectively and to ask questions and make clarifications throughout the data collection process, thus providing an important and innovative revision to my research protocol. These questions and clarifications can be said to reflect several elements of what Leah Lakshmi Piepzna-Samarasinha (2018) describes as "crip emotional intelligence," or "skills we [disabled communities] use within our own cultures and with each other" (69). As Piepzna-Samarasinha (2018) elaborates, crip emotional intelligence includes "figuring out how to communicate using smaller words, not academic words, different words than you just tried, writing. It's waiting for someone to be done finishing spelling out a sentence on their augmented communication device before responding. It's using text. It is not assuming that audist and academic ways of communicating are the smartest or the best" (70). In the case of Eduardo and the other men in his booth, participants facilitated language access through crip emotional intelligence by innovating a method for taking a survey that then resulted in research participants who were more informed about the questions they were answering and the research to which they contributed. As Christa Teston and her coauthors (2019) explain, "The design of surveys is a fundamentally rhetorical act," one that should always account for research participants with many precarious positionalities, including various dis/abilities, races, and linguistic histories (321). An intersectional, interdependent attunement to research methodologies can foster innovative practices grounded in participants' expertise and experiences, as evidenced by Eduardo's adaptation of the survey method in collaboration with the other men who took the survey in his booth.

In telling the story of my data collection at the Bronco, I'm hoping to illustrate several things. First, alongside many disability studies scholars, I believe no research method or project is fully accessible; we as researchers always bring our biases and perceptions to any research protocol, and we thus make mistakes and assumptions about our participants and their engagement. While these mistakes are common, our accessibility failures are not always inconsequential, since, as critical race scholars illustrate, our engagement with other human beings can have both visible and invisible impacts. At the same time, I share the story of my day at the Bronco as well as multiple other stories throughout this book "as opportunities to reflect deeply on the beauty, complexity, and pain of research" (Price and Kerschbaum 2016, 22). I want to document both my mistakes and my successes as a researcher as I extend implications and possibilities for others seeking to practice justice-driven multilingual technical communication. To me, this requires critical reflection as well as research-driven propositions about how we as technical communicators can continue working to do better to not cause harm and work toward justice in our collective practice. Finally, I share this story as a grounding example of how I've continued to theorize intersectional, interdependent methodologies alongside collaborators, communities, and participants I'll be introducing in the case studies presented in this book.

## DEFINING INTERSECTIONAL, INTERDEPENDENT METHODOLOGIES

I define the methodology I use in this book as "intersectional and interdependent" (see Gonzales 2018; Gonzales and Butler 2020). I believe that multilingual research requires an attunement to interlocking oppressions and accessibility; and intersectional, interdependent methodologies, drawing from work in both critical race studies and critical disability studies, can help technical communication researchers make more informed methodological decisions when working with multilingual communities (Collins 2019; Pimentel 2008; Medina 2014; Piepzna-Samarasinha 2018). Through intersectional, interdependent orientations, technical communication researchers can move away from the "damage-based" perceptions of multilingual communities—those that position speakers of non-normative languages (e.g., standard white English) as less capable, less articulate, less intellectual than the norm. This does not mean I advocate for the positioning of multilingualism as a disability. Rather, intersectional, interdependent methodologies recognize language difference as a reality in all communicative contexts and centralize the expertise of marginalized communities, including disabled communities, as critical to successful language access.

### *Intersectionality*

I recognize that using the term *intersectionality*, particularly as a white Latina, is highly contested, especially when the term is used without being clearly rooted in the work of Black womanist scholars and in the experiences of Black women (brown 2017; Crenshaw 1989; Collins 2019; Lorde 2012). As Black feminist and (dis)ability studies scholar Sami Schalk (2018) explains, "Intersectionality is . . . too often assumed to only apply to minority identity positions or [is] understood as a purely additive and ever-expanding term" (8). When choosing to use intersectionality as a methodological framework for engaging with multilingualism, I do so because I agree with Black womanist scholars that the freedom of Black women should be at the heart of all antiracist, anti-oppression, decolonial projects. As the Combahee River Collective (1977) explains in its manifesto, "If Black women were free, it would mean that everyone else would have to be free since our freedom would necessitate the destruction of all the systems of oppression" (215). The destruction of all systems of oppression is the type of social justice that technical communicators should work toward and that begins with an interrogation of anti-Black racism and its influence on *all* our work.

I also deeply respect and align with Patricia Hill Collins's (2019) illustration of the potential for intersectionality to be expanded and applied as a critical social theory. In *Intersectionality as Critical Social Theory*, Collins (2019) discusses the dimensions of intersectionality and its multiple core constructs. She explains that "intersectional knowledge projects achieve greater levels of complexity because they are iterative and interactional, always examining the connections among seemingly distinctive categories of analysis" (47). At the same time, Collins (2019) warns that researchers who embrace intersectional methodologies should not do so uncritically, as claiming to use an intersectional methodology does not mean you are actually engaging in a social justice project (i.e., you can enact an intersectional project that causes harm to people across social identities). Instead, Collins (2019) outlines three dimensions of critical thinking through which intersectionality is taken up: as a metaphor, as a heuristic, or as a paradigm. It's important for researchers using intersectional methodologies to clearly explain how they are uptaking this methodology and how they are grounding their work in intersectionality's core constructs, which include relationality, power, social inequality, social context, complexity, and social justice (49).

I use intersectionality as a methodology in this book for multiple reasons. First, intersectionality is grounded in social movements and was developed by Black feminists working to redress social injustices in the world (Combahee River Collective 1977; Crenshaw 1989). As such, intersectionality is applicable in the academy, but it is not bound by the academy (Collins 2019). Instead, intersectionality as a methodology can be used to foster "dialogical engagement" across "communities of inquiry" (15). When embedded as the foundation of a project focused on participatory design, as I seek to do in this book, intersectionality can also foster social justice–driven innovation.

As Collins (2019) explains, intersectionality is often used to describe multiple interlocking oppressions across and through social categories such as "gender, race, ethnicity, nation, sexuality, ability, and age" (10). What is less often highlighted, however, is the fact that just as intersectionality can be used to understand interlocking layers of oppression, it can also be used to understand interlocking layers of *resistance*, since "these terms also reference important resistant knowledge traditions among subordinated peoples who oppose the social inequalities and social injustices that they experience" (10). Thus, just as designers, scholars, and practitioners can look to intersectionality to begin to understand how oppression works and how racism is perpetuated, we can also look to intersectionality to better understand how multiply marginalized people resist oppression and innovate solutions for and with their own communities. We can practice intersectional methodologies to develop mechanisms for redressing oppression alongside other marginalized people (Piepzna-Samarasinha 2018).

Intersectionality as a methodology for understanding innovation requires an attunement not only to how individual people design tools and technologies for their own purposes but also to the way collective groups of people leverage their experiences to design tools and technologies that are useful and accessible across social categories. Because language is best understood through intersectional lenses that account for interlocking relationships among social categories (e.g., race, ethnicity, gender), intersectionality can also help researchers understand the multiple intertwined layers encompassed in a multilingual experience.

Because language is always connected to power, I align with African and African American language scholars who consistently demonstrate that language and race are co-constructed and thus that scholars who want to analyze language diversity should always acknowledge the interlocking connections among race, language, and power. Intersectionality, then, as "a lens through which you can see where power comes [from] and collides, where it interlocks and intersects," provides an opportunity to study language in constant relation to multiple forms of power, innovation, and oppression—not allowing researchers to position language as a neutral by-product of communication (Crenshaw 2017). Further, intersectionality, through its emphasis on interlocking rather than discreet forms of power and oppression, can help researchers understand the fluidity of language and its constant adaptations based on cultural-rhetorical contexts. As Ofelia García and Li Wei (2015) illustrate through their discussions of "translanguaging," languages are not discreet, stable categories or sets of communicative resources; instead, languages move and change and are adapted by speakers in each interaction, based on several interrelated factors connected to race and privilege—including the racial dynamics of a particular community, the rhetorical exigence of a specific utterance, or just a speaker's preference in a particular moment or instance. Thus, from a linguistic perspective, intersectionality helps researchers understand language and its connection to power on dynamic and interlocking levels that should not be abstracted and decontextualized (Alim, Rickford, and Ball 2016). Furthermore, intersectional approaches to technical communication are critical as the field works toward engaging with access efforts that consider how all "systems of oppression are locked up tight," since "ableism helps make racism, Christian supremacy, sexism, and queer- and transphobia possible" (Piepzna-Samarasinha 2018, 22). Thus, to understand how language is in/accessible, we need methodologies that foreground the expertise of people with interlocking layers of experience who can innovate strategies for combating oppression on multiple levels simultaneously.

Through an intersectional methodology, then, technical communication researchers cannot simply ignore the presence of race in a research project, omitting or positioning discussions of race as those that are only relevant to racialized subjects (Williams and Pimentel 2014). Instead, as Schalk (2018) elaborates, intersectionality is "an epistemological orientation and practice that is invested in coalition building and resistance to dominant structures of power" (8). Intersectionality, then, can help technical communicators recognize, rather than ignore, "the role that language plays in shaping our ideas about race and vice versa" (Alim, Rickford, and Ball 2016, 1). Ultimately, my goal in centering intersectionality as a methodology in multilingual technical communication is to insist that language diversity projects that ignore race can very easily erase important nuances about the role language plays in technological innovation while also maintaining oppressive power structures that deem particular (i.e., white/Western) people as the only ones capable of being innovative and intelligent. Thus, embracing intersectionality in language research, both within and beyond technical communication, requires an interrogation and intentional redressing of anti-Black and anti-Indigenous racism that has historically and contemporarily structured power relations in research (Tuck and Yang 2012; Alim and Pennycook 2007; Rosa and Flores 2017). Along with Schalk (2018), "I [too] am personally still invested in the potential of intersectionality and I find power in its particular women-of-color lineage even as I am aware and critical of how it has been used in limiting, static, and even regressive ways" (8).

### *Interdependency*

While intersectionality as methodology provides a framework for understanding multiple forms of oppression and their connections to language and power, interdependency, particularly as it is presented by critical disability studies scholars, helps articulate the shared responsibility of engaging language difference through intersectional perspectives (Piepzna-Samarasinha 2018). Too often, issues of language accessibility, like disability considerations, are dismissed as inconsequential in technical communication research, especially when the focus is on supporting globalization (Haas and Eble 2018). In many of these scenarios, translation or multilingual engagement is outsourced to third parties who "take care" of the "language problem" in global technical communication research. Apart from being oppressive and placing undue labor on the most marginalized members of our work (i.e., multilingual communities and multilingual communities of color), this type of detachment from engaging with the dynamics of language diversity also prevents researchers from leveraging the potential insights language diversity can bring to our research projects and research findings. Interdependency, as defined by Julie Jung (2014), is an "ethic for intellectual work" in which all people involved in an interaction take responsibility for making communication accessible in a shared space (101). Interdependency is a product of the human condition in which we all rely on other human beings in various ways through different relationalities, including but not limited to language (Gonzales and Butler 2020, n.p.). As Kerschbaum (2014) explains, communication is a "rhetorical performance," in which all individuals should "acknowledge their responsibilities to others in communication, . . . maintain attentiveness in varied ways of interacting, and . . . cultivate openness to interactional possibility" (118). In this way, as Price and Kerschbaum (2016) have also noted, interdependency can provide a way for researchers to interrogate how communicative acts, as rhetorical performances, rely on a collective negotiation of access among all parties involved in a discussion.

Interdependency and intersectionality have long been connected in "care webs" or collectives designed by queer, trans, disabled, Black, and Indigenous peoples and other people of color (Piepzna-Samarasinha 2018, 66). For example, intersectionality is listed as the first principle of Disability Justice, "a political movement and many interlocking communities where disability is not defined in white terms, or male terms, or straight terms" (Piepzna-Samarasinha 2018, 22). As Patty Berne, Aurora Levins Morales, David Langstaff, and Sins Invalid (2018) explain in the "Ten Principles of Disability Justice," "We know that each person has multiple identities, and that each identity can be a site of privilege or oppression. The mechanical workings of oppression and how they output shift depending upon the characteristics of any given institutional or interpersonal interaction; the very understanding of disability experience itself is being shaped by race, gender, class, gender expression, historical moment, relationship to colonization, and more" (228). Thus, as I propose that intersectional, interdependent methodologies be embraced in technical communication, I also recognize that these methodologies have long been connected, embodied, lived, and practiced by disabled communities of color.

Through interdependent approaches to multilingual research, technical communicators can position language access within the broader context of intersectional access that considers multiple interlocking identities and the languages that facilitate communication among groups of people. Intersectionality and interdependency can help support what Collins (2019) describes as "flexible solidarity *within* a particular political community and fostering solidarity among/across political communities" (171, original emphasis). Thus, intersectional, interdependent methodologies position translation and interpretation as part of broader accessibility networks that consider how researchers, participants, and the environments and communities in which we do our work interact and negotiate access and power in every interaction. As I will continue to demonstrate, language access, and accessibility more broadly, are shared responsibilities that cannot and should not be outsourced or retrofitted in technical communication practice. Rather than being ignored or placed as an afterthought to research design, an intersectional, interdependent approach to designing multilingual experiences demonstrates that translation can be embraced through interdependent orientations that foster "broader identification processes" that highlight "the lived experiences that bring differences alive" (Kerschbaum 2014, 9). Disability studies frameworks of interdependency, as Price and Kerschbaum clarify, can result in research methodologies that center "care, commitment, and acting with others in mutually-dependent relationships," where relying on others to access information is not a matter of choice but an intentional, necessary practice (27). Because accessibility makes all users' experiences more effective, all users, participants, and stakeholders involved in research interactions should share the responsibility of providing access, including linguistic access and all its intersectional implications.

## APPLYING INTERSECTIONAL, INTERDEPENDENT METHODOLOGIES IN TECHNICAL COMMUNICATION

Following the work of queer/trans Black, Indigenous, people of color, and disability activists, I bring intersectionality and interdependency together as methodological frameworks for the projects presented in this book, both because I think both frameworks are independently useful and because interdependency and intersectionality function more effectively in relation to one another. Many critical disability studies scholars bring attention to the necessity of engaging more deeply with issues of race, intersectionality in particular, within disability studies. Anastasia Liasidou (2013), for example, explains that "disablism forms part of an intricate web of social conditions that subjugate certain forms of 'student-subjects' and create compounding forms of oppression and exclusion that need to be addressed through relevant education policy and practice" (299). Other scholars, such as Schalk (2018), point to the value critical race scholars can both bring to and find in disability studies, explaining that "while black feminist theories have done so much to demonstrate the relationship of various oppressions, (dis)ability is rarely accounted for in black feminist theory" (3). At the same time, Schalk (2018) explains that "disability studies scholars have generally not recognized black feminist work on health activism, illness, and access to medical care as properly disability studies" (4).

While Schalk (2018) acknowledges that (dis)ability is often engaged, though not named, in Black feminist research, she argues that more conversations between Black feminist and disability studies scholars can expand methods and methodologies for engaging productively with difference across and beyond specific identifications. As disability activist Piepzna-Samarasinha (2018) further clarifies:

> When we do disability justice work, it becomes impossible to look at disability and not examine how colonialism created it. It becomes a priority to look at Indigenous ways of perceiving and understanding disability, for example. It becomes a space where we see that disability is all up in Black brown/queer and trans communities—from Henrietta Lacks to Harriet Tubman, from the Black Panther Party's active support for disabled organizers' two-month occupation of the Department of Vocational Rehabilitation to force the passage of Section 504, the law mandating disabled access to public spaces and transportation[,] to the chronic illness and disability stories of second wave queer feminists of color like Sylvia Rivera, June Jordan, Gloria Anzaldúa, Audre Lorde, Marsha P. Johnson, and Barbara Cameron, whose lives are marked by bodily difference, trauma-surviving brilliance, and chronic illness but who mostly used the term "disabled" to refer to themselves. (22–23)

When disability "is not defined in white terms" (Piepzna-Samarasinha 2018, 23), it opens up possibilities to engage with complexity in the human experience, to recognize how the "bodymind" (Price 2011; Schalk 2018) can encompass both physical and mental disabilities that are intrinsically tied to oppression and that are navigated intergenerationally, trans-disciplinarily, and across communities by Black and Indigenous people of color who have long been taught to hide their needs and desires to conform to a white, colonial heteropatriarchy.

As a language and translation scholar who has both experienced and continues to witness the anxiety, trauma, and embodied pain that can go into language work in context—as mothers fight to seek and understand healthcare for their families, as families seek employment and documentation, as under-resourced communities continue to be the most heavily impacted by health disparities and racist legal systems—I find intersectionality and interdependency, and the connections between critical race theory in language and disability studies, to be critical components of how technical communicators should (re)image and (re)engage with multilingualism in contemporary contexts. Rather than ignoring the trauma, anxiety, pain, and joy language transformation brings to marginalized communities, particularly those that have to engage in high-stakes processes (e.g., health and legal proceedings) in standardized languages like mainstream white English, technical communication researchers should recognize the ways multilingual experiences shape individuals and influence the contexts in which research happens. When multilingual communicators use their language practices in formal environments under standardized protocols like research, we bring with us our previous language-learning experiences, our bodyminds and language skills, as well as our various visible and invisible dis/abilities. Intersectional and interdependent methodologies can thus help technical communication researchers account for these multiple layers of experiences embedded into multilingual communication.

In the case studies presented in chapters 4–6, I describe how intersectional, interdependent methodologies helped me co-develop opportunities for engaging with multilingual communities as we collaboratively designed information, tools, and resources in multiple languages. In chapter 4, I provide an overview of the specific methods I engaged in at each research site. Embracing an intersectional, interdependent methodology in these studies meant that I thought about how to incorporate multiple entry points into what "participation" might look like for different participants in different contexts at different points in time. Rather than follow a standardized research protocol, an intersectional, interdependent methodology meant that I adapted interview and focus group protocols based on the bodyminds sharing space with me and each other on a particular day.

For example, rather than simply handing out usability scenarios for participants to complete, enacting an intersectional, interdependent approach to usability studies meant that I paused to co-define and co-translate specific terms on a scenario or other prompt so participants in each research space could ask questions about the terminology. An intersectional, interdependent methodology means that no term is assumed to be "default" or "standard" and that no user in a research context should be assumed to define terms in the same way as the researcher. As such, intersectional, interdependent methods stem from long-standing practices developed by multilingual disabled communities of color, for which language "access" has always considered intersecting positionalities to be avenues for collaborative innovation. To innovate in community within multilingual contexts requires an attunement to the embodied difference present in every research space and the flexibility to adapt or rethink research protocols based on how participants are feeling in the group, the environment in which the research is taking place, the historical underpinnings and racial and linguistic dynamics of a group, and the individual dispositions of each group member in conversation.

As I demonstrate in chapter 3, an intersectional, interdependent methodology can be applied to various methods already common in technical communication and user experience research. The difference in this approach lies in fostering open participation by moving away from standardized expectations about what communication should look like in research spaces. While not always successful in making any research space fully accessible, an intersectional, interdependent methodology provides the impetus to pause in our research practice and make space for dialogue through spoken, written, gestural, visual, and/or otherwise embodied multilingual praxis. If we as technical communication researchers want to account for and incorporate diverse perspectives into our designs, then we have to be prepared to sit with the discomfort that comes with not being able to understand every utterance made in our research contexts. We have to understand that we'll need to reframe and revise our questions alongside participants and consider how our very presence in a research space inherently shifts what we might deem as our "findings." These considerations are the foundation of a multilingual experience, which can be best understood through the intersectional, interdependent frameworks introduced in this chapter.

## A NOTE ABOUT DISCLOSURE

Following the work of disability studies scholars who clarify that in communities of color, long-standing histories of oppression and racism sometimes prevent disabilities from being recognized and named and in agreement with Piepzna-Shamarasinha (2018), Schalk (2018) Price (2011), and Christina V. Cedillo (2018), who recognize that disabilities can be un-diagnosable and invisible as defined through Western paradigms, I did not ask my participants to disclose their disabilities during our project. However, through an intersectional, interdependent methodology, I sought to establish space in our conversations for disabilities to be part of our collaborative discussions, and I also shared my own experiences of linguistic trauma and anxiety when discussing translation and interpretation with various participants. As part of our participatory translation activities and conversations, participants also discussed their access needs and preferences, which frequently opened space for further discussion about individuals' dis/abilities in the context of multilingual design (Butler 2017). As Janine Butler (2017) explains, "Our bodies are being constructed via interaction with audiences, akin to the way our compositions are always co-constructed by readers" (n.p.). As such, paying attention to embodied difference in research spaces can foster different modes of interaction, including but not limited to translation. Through this perspective, what I hope to illustrate in the projects described in chapters 4–6 is that discussions about access and dis/ability should always be incorporated into the design and sustainability of multilingual experiences in technical communication and should not be reserved for circumstances where disabilities are disclosed by participants up front. In this way, as disability studies scholars have long argued, access can be made (more) possible for all members of a research community.

## CONCLUSION

Social justice–driven scholars in technical communication continue to call for more reciprocal collaborations among multiply marginalized scholars who focus on issues of diversity in and beyond technical communication (Walton, Moore, and Jones 2019; Haas and Eble 2018). Recent work in technical communication, for example, highlights the value of expanding the field through justice-driven "coalitions" (Jones 2020; Walton, Moore, and Jones 2019) that engage social justice from multiple perspectives (Haas and Eble 2018; Agboka and Matveeva 2018). This expansive approach is critical to the sustainability of social justice initiatives that work actively to redress oppression on multiple levels and that seek to engage multiple stakeholders in the labor of doing social justice work (Walton, Moore, and Jones 2019). As Kerschbaum (2014) has noted, "Attention to marking difference, when performed in conjunction with attention to various identification processes, can help us mediate between broad conceptual tools for talking about difference and the unique qualities of individual moments of interaction" (7).

These calls for building "coalitions for action" in technical communication are what inspired me to propose intersectional, interdependent methodologies as a way to (re)conceptualize technical communication's engagement with language diversity. I've often been asked if technical communication researchers who want to work toward social justice "should" or "shouldn't" do research in communities whose languages vary from those of the researcher. Intersectional, interdependent methodologies can help answer this question, specifically by helping technical communication researchers understand how power is always at play in multilingual interactions, whether it is acknowledged or addressed or not. Our field's best practices for engaging in research with marginalized communities can take on an additional level of rigor and complexity when language diversity is centralized as a critical component of global technical communication research. This type of reframing, in my opinion, can benefit from intersectional, interdependent methodologies that recognize how (1) race and racism influence all human interaction; (2) power and positionality operate on interlocking levels of identity, marginalization, and privilege; and (3) access can be best achieved through the shared labor of all stakeholders involved in an interaction. Furthermore, putting intersectionality and interdependence in conversation with multilingual technical communication research can help researchers expand the way we engage with communities for which identity, privilege, and marginalization are complexly intertwined.

For example, recent work in disability studies has pointed to the importance of recognizing that (dis)ability is a fluid identity that extends far beyond "discrete, self-evident categories" such as "disabled" and "nondisabled" (Schalk 2018, 9). Recent work on what can be categorized as "invisible" or "undiagnosed" disabilities points researchers to the importance of expanding their attention to "non-normative" bodies in all their forms and iterations (Cedillo 2018; Price 2011). This is particularly important when working within/in communities of color, for which trauma and racism permeate many other identity categories, including but not limited to race and language. At the core, embracing an intersectional, interdependent methodology can push technical communication researchers to recognize and perhaps dismiss misguided assumptions about the bodymind(s) present in our research, our communities, and our classrooms (Price 2011). Through interdependent, intersectional methodologies, we can orient to language from a foundation of difference as fact rather than difference as exception in our research praxis. The key to practicing an intersectional, interdependent methodology, then, is in pairing this orientation with methods that foster participation from the bodies, languages, and perspectives that are "too often absent" from our scholarship (Haas 2012).

# 3
# RESEARCH DESIGN

In describing "designing multilingual experiences in technical communication," I reference several methods and concepts that are already familiar to technical communication researchers and practitioners—including participatory design, user experience, and multilingualism. By putting these concepts together, I emphasize that "multilingualism" encompasses much more than words in different discreet languages. Instead, multilingualism is an *experience* that is directly tied to individual and communal histories, embodied practices, words, sounds, and gestures (Gonzales 2018). This expansive view of multilingualism echoes the "culture of access" proposed by Ada Hubrig and coauthors (2020), who emphasize that access is "the dynamic, collective movement of creating spaces where multiple marginalized disabled people with a wide range of needs can engage in whatever manners they choose" (91). Thus, what I propose in this book is that as technical communicators continue to engage with multilingual research, we should pay more attention to *all* elements of multilingual experience, including but not limited to those alphabetic sounds and words that are too often centralized in discussions about language diversity (Agboka 2013). As I demonstrate in this chapter, through an intersectional, interdependent orientation, common methods in technical communication and information design can be expanded to encompass broader multilingual contexts, communities, and spaces.

In chapters 4–6, I will introduce localized case studies that illustrate how intersectional, interdependent methodologies can help technical communication researchers further interrogate, and therefore be prepared to design, multilingual experiences that benefit rather than harm marginalized communities. Each project described in this book encompasses multiple moving parts and individual research questions. For the purposes of the data stories I share in chapters 4–6, the overarching questions I asked were (1) what does technical communication look like in multilingual contexts, particularly when approached through an intersectional, interdependent methodology; and (2) how can technical

https://doi.org/10.7330/9781646422760.c003

communicators design multilingual experiences that benefit (rather than ignore or harm) linguistically and ethnically diverse communities in both local and global contexts?

While the individual projects also encompass different methods that align with the contexts of a particular community, there are specific methods that I used and adapted across contexts to answer my overarching questions. These methods include participatory translation, user experience methods (e.g., card sorting, usability testing, affinity diagramming, journey mapping), and human-centered design, as well as focus groups, surveys, and interviews. In the sections that follow, I provide an overview of these methods.

## PARTICIPATORY TRANSLATION

In describing decolonial methodologies for social justice, Godwin Y. Agboka (2014) argues that to do decolonial research in global contexts, technical communicators need to develop new methods that span beyond the colonial project. As Agboka (2014) argues, "Intercultural professional communication research needs to develop a coherent body of new methodologies with their corresponding methods that are cognizant of local logics, rhetorics, histories, philosophies, and politics. By doing that, our research approaches will answer the call of social justice, which hinges on reflexivity, liberation, and empowerment" (298).

In another project, Agboka (2013) proposes "participatory localization" as a social justice methodology for conducting research in un-enfranchised/disenfranchised cultural sites. Explaining that "the social justice implications of technical communication localization, particularly in nonindustrialized, developing, and unenfranchised or disenfranchised cultural sites[,] have yet to be critically examined" (29), Agboka (2013) calls for a "more intuitive and user-sensitive localization approach that is reflective of sociopolitical issues existing at the user's site—an approach that will be undertaken from the ground up" (42). In technical communication scholarship, localization has long been highlighted as an alternative or companion to translation—as the practice that goes beyond language issues when adapting tools and technologies for specific contexts (Agboka 2013; Batova and Clark 2015; Sun 2012). In describing participatory localization, Agboka (2013) argues that "when the focus is only on linguacultural factors, significant issues such as local knowledge systems, political issues, economic implications, and legal systems prevailing at users' sites during the localization process are overlooked, if not ignored" (29). While Agboka (2013) and other technical communication scholars argue for the value of localization and participatory or "user-localization" (Sun 2012), in this project, I bring the focus back to language transformation through what I call participatory translation. I choose to emphasize language transformation not to dismiss the importance of broader localization but rather to highlight how language can function as a method for encouraging participation and transformative engagement by historically marginalized, often ignored, and disenfranchised publics.

In professional translation scholarship, researchers and teachers frequently point to the fact that being a professional translator or interpreter is not the same as just being a bilingual or multilingual communicator. Professional translation and interpretation require specific training in facilitating communication among various parties from different linguistic and cultural backgrounds (Gonzales and Bloom-Pojar 2018). However, bilingual and multilingual individuals, through their linguistic, cultural lived experiences, do carry expertise that can be of value to technical communicators and information designers, particularly when working in multilingual global contexts. In the projects I introduce in this book, I find participatory translation, rooted in the democratic participation goals of participatory design (Racadio, Rose, and Kolko 2014), to be a helpful entryway into inspiring engagement from participants who may not typically see themselves as experts in common design scenarios. As W. Michele Simmons (2008) explains in her discussion of civic engagement in policymaking, "all participation is not equal," particularly if "there is an unequal distribution of power or privilege granted to particular groups" (13). As some technical communication researchers have pointed out in reference to participatory design activities, the notion of "bringing everyone to the table" to participate in a research and design project does not mean that every individual will have equal say in the conversation or will feel comfortable (often due to racist and oppressive histories working within white/Western/ableist design contexts) sharing their perspectives (Rose and Cardinal 2018; Cardinal, Gonzales, and Rose 2020; Price and Kerschbaum 2016; Sackey 2020). This is especially the case when considering participation from the perspective of disabled communities of color who are often made to feel like "expensive burdens" who "detract" from academic experiences (Hubrig et al. 2020, 88).

Robert Racadio, Emma J. Rose, and Beth E. Kolko (2014) explain that while participatory design (PD) has been adapted for various different contexts, "as PD practices become more widely adopted, there is a risk that its founding political values will come [to be] taken for granted, ignored, or overlooked," particularly when PD is applied to projects with marginalized communities (49). Racadio, Rose, and Kolko (2014) explain that marginalized communities "often possess limited technical expertise and lack the material means to successfully implement [a PD] project" (49). However, marginalized communities, particularly multilingual communities of color, do possess extensive linguistic and cultural expertise that can and should inform technology design and innovation. Participatory design is thus intended to bring this linguistic expertise to the forefront of design activities with an attunement to linguistic and embodied difference.

I incorporated translation activities into different elements of the projects I describe in this book to provide multiple different avenues for participation and engagement and to highlight, rather than ignore or erase, the role language diversity plays in multilingual research in global contexts. As I will illustrate in chapters 4–6, when technical communication researchers work with multilingual communities, the labor of language translation and interpretation is often segmented or ignored while emphasis is placed on the "real" work of the project—as something that can either be taken care of by a single individual (i.e., a translator or interpreter) or be done after the project is completed (e.g., translating material into other languages after it is created in English). Instead of following these models, participatory translation requires that an entire group—facilitators, participants, observers, and others—engage with the process of language transformation in whatever capacity they might deem possible or necessary. For example, in an interview with dscout, UX designer and sociologist Alba Villamil (2020) explains the need to reframe and expand understanding regarding common research processes and protocols when working with marginalized communities. Using the example of the research consent process as something that demands further interrogation in work with communities, Villamil (2020) asks: "What if in our interactions with communities we reframe consent as something they can strategically withhold rather than something they sign away? In practice, this could take many forms. We could draft granular consent forms. We could design data repositories where research participants could delete their data ad hoc. We could work with community leaders and host townhalls about consent and data privacy. We could even develop accountability boards" (n.p.).

Participatory translation, as I practice it alongside communities in this book, is a method for engaging in the type of rhetorical reframing Villamil (2020) advocates for. To reframe concepts that are typically embraced or accepted without questioning (e.g., consent), it's important to open up conversations (broadly defined) with various stakeholders about how a single concept can be interpreted by different groups, in different languages, and/or through different racial, gendered, and embodied positionalities. Participatory translation provides an avenue for groups to collectively decide and discuss, rather than just assume, how words or concepts are to be taken up in a specific project and thus how access will be fostered through relationality (Hubrig et al. 2020). For example, to engage in participatory translation activities around the notion of consent, a facilitator may write and verbalize the word *consent* alongside representative images, sounds, and/or descriptions on a whiteboard and ask participants to come up with as many possible definitions, translations, images, sounds, examples, or interpretations to represent the term as they can. In this way, "the" definition of consent is expanded into many definitions, often with translations, as participants can be welcomed to also discuss consent in various languages. Through a participatory discussion focused on how the word *consent* itself can be translated (both across languages and across various positionalities and perspectives), the group can then decide which approaches to consent are most suitable for a specific project in a local context. A participatory translation approach thus echoes fluid notions of language that have long been established in sociolinguistic research, where researchers acknowledge that there isn't, for example, a "single" Spanish but that there are multiple versions of Spanish consistently used in Spanish-speaking communities. Furthermore, participatory translation embraces an interdependent approach to research praxis by providing opportunities for members of the group to pause and collectively decide how certain words, concepts, and activities will be made accessible to everyone present.

It's important to acknowledge that, like participatory design, participatory translation activities don't automatically guarantee equal participation from all members of a group. This is especially the case in situations where white facilitators are working with communities of color, where communities continue to experience discrimination and oppression. Alongside encouraging all members of a group to provide possible translations or interpretations of concepts to be used in a project, engaging in participatory translation activities through intersectional, interdependent methodologies requires a reframing of language access and clear or effective communication.

When engaging in participatory translation activities, it's important for researchers to leave space for their own linguistic limitations to influence what and how they will engage with the conversations taking place in the group. Opening up research protocols into what Rachel

Bloom-Pojar (2018) describes as a "translation space," or any context that encompasses the negotiation of meaning among different stakeholders, means that language is negotiated and meaning is not assumed and that not all members of the group, including the researcher(s) or facilitator(s), will understand every word uttered in an interaction. In participatory translation activities, participants may engage in conversations that do not include the facilitator, particularly as they decide how they would like to partake (or not) in a particular discussion. Through intersectional, interdependent methodologies, the role of the researcher or facilitator in this case is not to "bring back" the discussion to a common language that can be understood by all but rather to find nuance in the complexity of the conversation, pay attention to non-verbal cues and aspects of the conversation, and find alternate forms of listening and engaging until their perspective is once again welcomed (if it is welcomed at all). As Hubrig and coauthors (2020) explain, "So often, access is described as a vertical framework: access is given by someone with more power to someone with less power" (92). Contrastingly, participatory translation can help reshape what access means in a shared research space, where nobody is guaranteed access to all the content being shared and where a shared understanding of the conversation is reliant on the interdependent connections made between the group and the trust and relationality established in the research space. If trust and relationality are missing, access will be limited to those who are already part of the established community. As Patricia Hill Collins (2019) explains, the concept of relationality "shifts focus away from the essential qualities that seemingly lie in the center of categories and toward the relational processes that connect them" (45). Thus, intersectionality and interdependency, through participatory translation, can shift the focus of a project from a single language to the relationality among people and their many languages and how they interact in a shared research space. The type of reframing and reorientation participatory translation provides, through an emphasis on language as expertise, can lead to effective multilingual experiences in and beyond technical communication.

## USER EXPERIENCE AND HUMAN-CENTERED DESIGN

In referencing "multilingual experiences," I also signal the user experience methods and approaches to interaction I employ in this book. User experience and technical communication have a long, intertwined history focused on engaging various stakeholders in designing accessible information (Redish and Barnum 2011). User experience can be defined as "a person's perceptions and responses resulting from the use and/or anticipated use of a product, system, or service" (ISO 9241–210, quoted in Rose et al. 2017, 214). Through this perspective, user experience research is positioned as a corporate-driven enterprise that extracts knowledge and feedback from people for the benefit of capitalist gain (Cardinal, Gonzales, and Rose 2020). As Alison Cardinal, Laura Gonzales, and Emma Rose (2020) argue,

> UX design, as it is practiced in industry and popularized by technology companies, who rely heavily on design to monetize their products and services, tends to engage users in design, but at arm's length. UX practitioners use a variety of methods to understand the people they are designing for and [to] uncover their needs, including surveys, interviews, ethnography, A/B testing, eye tracking, and analytics. These practices are deeply embedded into how UX research is practiced in industry and often taken for granted. Users and their expertise are seen as valuable and that knowledge is extracted solely for the good of the product or system, often without reciprocation beyond a small stipend or token of appreciation. (2)

UX research can cause harm to marginalized communities that are already positioned in deficit perspectives and constantly exploited in research. At the same time, as Rose and her colleagues (2017) also argue, "UX practices such as audience and task analysis, technical writing, and usability testing can help nonprofit organizations [and, I would add, other groups and initiatives led by and for marginalized communities] pursue their missions" (214). In the projects I describe in this book, I incorporate several UX activities intended to help generate feedback from communities and inspire innovation alongside intersecting goals. These include various mapping exercises intended to help groups theme multiple ideas into common threads or patterns (e.g., affinity diagramming, journey mapping); drawing, sketching, and gesturing activities where participants can design freely without needing to rely on digital tools and technologies (e.g., paper prototyping and low-fidelity prototypes); and usability testing, in which participants design a specific tool and then develop protocols for gathering feedback on their designs. My approach to engaging with UX research also follows human-centered design principles that emphasize human agency over products. According to ISO 9241–210:2010(E), human-centered design "enhances effectiveness and efficiency; improves human well-being, user satisfaction, accessibility[,] and sustainability; and counteracts possible adverse effects of use on human health, safety, and performance" (quoted in Rose, Björling, and Cakmak 2019, 604).

While I appreciate and embrace the emphasis on "human dignity and human rights" (Walton 2016) put forth by human-centered design principles and while I incorporate this approach into the projects represented in this book due to its value and emphasis on working with marginalized communities, I also recognize that "evidence of the extraction model of UX can be seen in the language and practices that surround much of UX design" (Cardinal, Gonzales, and Rose 2020). For example, the emphasis on "efficiency" and the assumption about what marks "improvements" in factors such as health and safety echo white/Western ableist values and orientations to satisfaction and well-being. For this reason, when employing white/Western/English–dominant methods in work with communities of color, particularly multilingual communities of color in global contexts, it's important to also trace how issues of privilege, power, and positionality (Jones, Moore, and Walton 2016) consistently influence design and research spaces. As Ann Shivers-McNair and Clarissa San Diego (2017) clarify in their discussion of "community strategy" as an approach to community-driven innovation foregrounded in inclusivity and collaboration, "the definitions of user, community, and diversity themselves must be continually localized in our work to engage across cultures and across theory and practice" (97). For this reason, I find it critical to pair UX research methods and human-centered design principles with participatory translation activities that allow for the type of (re)defining work Shivers-McNair and San Diego (2017) call for.

Through an intersectional, interdependent methodology, UX research becomes a method for interrogating what participation can or should look like in a shared space, for understanding who has access to particular methods of engagement, and for learning how power and positionality play into every research question, activity, or practice. In the projects I describe in chapters 4–6, for example, I incorporate UX research methods with frameworks that echo the local values and practices of the communities with whom I was fortunate to work. For example, in chapter 4, I describe how UX methods were adapted by youths who live and learn in the borderland city of El Paso, Texas. In this particular context, UX and participatory translation are combined with borderland theories and Chicanx sociolinguistics to open up conversations about how tools can be designed beyond stating language categories such as "English" *or* "Spanish," instead reflecting the constant mix of various Englishes and Spanishes local communities engage in within the borderland region.

In chapter 5, I pair participatory translation and UX activities with the history of colonization in South Asia to better understand how language practices in contemporary academic spaces in Kathmandu, Nepal, can still be guided by long-standing legacies of linguistic oppression while at the same time, multilingual researchers and community members in Nepal continue to find innovative solutions for navigating linguistically oppressive technologies. Finally, in chapter 6, I thread Indigenous epistemologies with decolonial theories alongside participatory design and UX to describe the work my research team in Oaxaca de Juárez, Mexico, is engaging with to bring visibility to the connections between language and land rights in and beyond the Mexican settler states. In short, embracing an intersectional, interdependent methodology when practicing UX research requires an interdisciplinary approach that acknowledges and centers participants' values and histories into design and innovation. As a field that has historically been interdisciplinary and focused on human interaction, UX has helped me better understand how technical communicators can recognize and support the "knowledge work" communities around the world are already engaging in (Grabill 2007). At the same time, as I demonstrate in the remaining chapters of this book, in applying UX methods and embracing human-centered design principles for this project, I was also pushed to consider how my own positionality and the positionalities of the languages used in each of the research contexts I introduce are both shaped by and can help reshape Western approaches to designing technical information and digital experiences for and with marginalized communities. As Simmons (2008) further clarifies in her discussion of participation in policy decisions, "encouraging citizens to contribute knowledge about how a policy will affect their community at the onset of a decision-making process is quite different from allowing citizens to respond to policies already determined" (13). Similarly, as I learned through the projects I describe in the remaining chapters, using UX methods to encourage participation can only be effective if participation is defined collectively and welcomed at every stage of the design process, ranging from the consent protocol to the resulting product, idea, or design that stems from a multilingual research project.

## FOCUS GROUPS, SURVEYS, AND INTERVIEWS

In the projects I describe in chapters 4–6, I work to make visible the way power relations shape interaction and how these interactions are grounded in long-standing histories of oppression and colonization. To do so, I provide snapshots into particular methods and instances throughout my engagement with various community groups, focusing on how language (and its inherent connections to power) already guides and can be used to more intentionally shape technical communication research.

In introducing various projects in a single book intended to describe multilingual experiences in technical communication, I emphasize that each experience, each project, each community calls for different methods and approaches to research that should always be localized for and with the participants who contribute their labor to this work. In the projects included in this book, I pair various theoretical frameworks with different methods at each research site, including more standardized research practices such as focus groups, surveys, and interviews. Since each project I describe is part of a larger constellation of initiatives intended to help researchers learn about multilingual experiences, certain pieces of each project are both included and excluded from the discussions deemed pertinent to this book's major arguments. In each chapter, as I introduce a specific method, such as a focus group discussion, survey protocol, or interview session, I aim to highlight not only the specific questions asked during a data collection period and the conversation that took place in that session but also the often ignored elements that influence these methods.

For example, in a typical focus group setting, participants are asked to engage in the discussion while also ensuring that they are not taking over the conversation, thus allowing other members of the focus group to share their perspectives. While this method can be fruitful as participants bounce ideas off each other and engage in a discussion, as with other UX methods and participatory design approaches, welcoming participation from multiple members of a group does not guarantee that equal participation will be achieved by all members of a group; thus, particularly for projects in which members of a group have different linguistic strengths, focus group methods should be paired with other approaches and intersectional, interdependent methodologies that recognize how power and positionality are impacting each conversation.

Similarly, incorporating individual interviews with participants can be a generative way to gain individual insights into how a design project is going and how group interactions could be improved. Yet in interviews, a researcher's positionality and embodied difference can influence the way participants respond and the perspectives participants feel comfortable sharing. As Simmons (2008) explains, traditional interview protocols need to be adjusted in civic and community contexts: "Technically complex public issues complicate the traditional notion of discourse because technical experts claim ownership of the technical issues and close off public debate even though these issues affect the public in very concrete ways" (10). Furthermore, as Margaret Price and Stephanie L. Kerschbaum (2016) explain, "When disability is assumed to be an important part of the qualitative interview situation (rather than something external that 'enters' the situation and then must be accommodated or compensated for), the interview's normative framework is both exposed and challenged" (20). As technical communication researchers, we can (perhaps unintentionally) close off discourse and conversation through individual interactions with stakeholders if we come into projects with assumptions about how participants should respond or participate. Thus, as Simmons (2008) elaborates, technical communicators need "a rhetoric for civic discourse" that works intentionally to "(1) identify and bring to the forefront unequal power relations that currently work to marginalize public involvement, (2) see the public as capable of contributing useful knowledge to the decision-making process, and (3) offer ways to include the public earlier and more significantly in the decision-making process" (10).

Table 3.1 provides a summary of the various methods I incorporated in each of the projects detailed in chapters 4–6. While I provide extensive detail on each of the contexts and communities with which I had the privilege to collaborate in each individual chapter, table 3.1 provides an overview of the different locations in which these projects took place: El Paso, Texas; Kathmandu, Nepal; and Oaxaca de Juárez, Mexico.

Within each research context, methods such as participatory translation, surveys, interviews, and focus groups were localized around the specific values and goals of the community participants. Through an intersectional, interdependent orientation, each method was adjusted to complement the scope of a single project as well as to fit within the specific affordances, desires, and constraints (in terms of resources, skills, and other elements) each context provided. For example, at times, interviews took place in our research space, at local coffee shops, or virtually depending on what was most accessible for participants. The timing and structure of each focus group or interview also shifted depending on how participants could best access the information presented on a particular day. Sometimes, rather than follow a protocol pre-set by me or other facilitators, the interview and focus group questions were developed by participants themselves. For example, in working with youths in an after-school program in El Paso, asking structured interview or focus group questions would not render useful responses. This echoes research on conducting usability testing with children, where researchers point to the importance of leveling power dynamics

Table 3.1. Methods used across research sites

| *Location* | *Participatory Translation* | *UX Methods* | *Human-Centered Design* | *Surveys* | *Interviews* | *Focus Groups* |
|---|---|---|---|---|---|---|
| El Paso/Ciudad Juárez | –Translation of a brochure related to diabetes treatment and prevention for Latino men<br>–Translation of a website and accompanying content for an after-school program with youths and their families | –Usability testing of a brochure and promotional materials for a diabetes treatment and prevention program<br>–Usability testing of various materials for an after-school program. Affinity diagramming, card sorting, journey mapping with youths and their families at an after-school program | –Emphasis on borderland theories and experiences, language fluidity, Chicanx sociolinguists | –Baseline demographic surveys for all participants<br>–Feedback surveys on experiences in the after-school program to assist in planning | –Interviews between mothers and their children at an after-school program | –Focus groups to engage in participatory translation of various materials, including a brochure, for a diabetes treatment and prevention program<br>–Bi-weekly focus groups with youths and their mothers to assess how the after-school program facilitation is going and what could be improved<br>–Focus groups with youths and their mothers in the after-school program to get feedback on the logo and website for the organization |

*continued on next page*

Table 3.1—*continued*

| *Location* | *Participatory Translation* | *UX Methods* | *Human-Centered Design* | *Surveys* | *Interviews* | *Focus Groups* |
|---|---|---|---|---|---|---|
| Kathmandu | –Participatory translation of scenarios and tasks for usability studies<br>–Participatory translation of websites related to Kathmandu<br>–Participatory translation of various guiding concepts for the group (e.g., digital, participation) | –Journey mapping, affinity diagramming, usability studies, personas and scenarios, content audit, landscape analysis | –Emphasis on how colonial histories in South Asia shape the role of English in academic institutions<br>–Acknowledgment and addressing of race and power relations among our group and in relation to caste system in Nepal<br>–Emphasis on participants' own academic goals and research interests rather than focusing on facilitators' goals | N/A | –Three individual video-recorded interviews with each participant to assess how the project is progressing and how students are understanding various concepts and ideas | –Focus group discussions focused on analyzing current digital representations of Nepali languages in online spaces |
| Oaxaca | –Participatory translation of various legal terms into Indigenous languages by various Indigenous communities<br>–Participatory translation of group concepts such as "participatory design" and "user experience" | –Journey mapping, affinity diagramming, usability studies, personas, and scenarios | –Decolonial approaches to participatory design<br>–Indigenous approaches to language, which connect language to bodies and land | –Initial survey from all participants to identify each participant's language history and home communities<br>–Feedback survey following our collaborative event | –Individual interviews with participants who discussed their journey as Indigenous language interpreters and translators | –Roundtable discussions themed around participants' interests<br>–Final focus group with all participants to assess our collaborative event and plan future iterations of the project |

when testing designs with youths (Als, Jensen, and Skov 2005). To this end, in our project we asked youths to prepare interview questions and interview each other or to work as focus group facilitators who could help us understand things such as what each participant liked to eat at home, what activities they liked to do when getting exercise, what types of dancing they enjoyed, and similar factors. At times, youths would interview their mothers about specific family recipes or histories. Through these activities, participants took ownership of the research elements of these projects and helped guide the direction of the team as a whole. This participatory approach was also embraced in the design of UX research activities such as usability testing, when participants wrote their own usability scenarios to test with each other and with other members of their communities. In this way, the overarching intersectional, interdependent methodology that guided the individual methods of this project fostered opportunities for flexibility and adjustment throughout the research process.

As described in table 3.1, a human-centered orientation to the methods described in each project also pushed me to engage with theoretical frameworks and additional methodologies that could help me better understand the various contexts in which my collaborators and I were working. These theories and orientations include borderland theories and Chicanx sociolinguistics, South Asian and postcolonial studies, and decolonial theories and Indigenous epistemologies. By centering these theories in technical communication projects, I emphasize that as technical communication researchers engage in work with multilingual communities of color, we should consider when and how we are applying white/Western frameworks to transnational and international research. For example, just because we apply UX methods to our research within a particular community without incorporating additional theoretical frameworks, it doesn't mean we are not embracing white/Western frameworks in our practice. For this reason, regardless of what methodological or theoretical orientations we embrace in our research as technical communicators, an attunement to methodological grounding is important and necessary. That is not to say that the frameworks and methods we have developed in technical communication research are not applicable in multilingual contexts. Instead, I advocate for added intentionality in the way we apply our technical communication methods and methodologies in community work and an openness to adjusting and expanding these methods and methodologies through the lessons we learn from our community partners and collaborators across the globe. This type of openness to adaptation and change is what has historically helped technical communicators succeed as stewards of organizations and is what can help the field continue to expand, question, and advance its social justice mission.

## DATA ANALYSIS

In each of the projects introduced in chapters 4–6, I will share individual stories and experiences to illustrate how multilingual experiences were created and sustained across contexts. While each chapter's structure varies, each one does include similar elements that I deem important to understanding the role language plays in collaborative, participatory research. These elements include (1) a historical discussion of the location in which each project took place, with a particular emphasis on the language relations of each specific location; (2) an introduction to the projects and participants discussed in each chapter; (3) a discussion of my own positionality in relation to these locations and to the participants of each study; (4) an illustration of what I call "language positionality," which requires me to acknowledge which languages(s) were used most frequently in each context, who benefited from the use of these languages, and how these languages shaped participation in a project; (5) specific grounded examples of how multilingual research took place in each context; and, perhaps most important, (6) narrative reflections written by participants at each research site after reading a draft of this book. In these reflections, participants were invited to read, reflect, and critique or make recommendations based on their own perceptions and analyses of this project. In addition, participants were encouraged to write their reflections in a language of their choice and to provide translations (or not) as they deemed appropriate. Thus, by incorporating narrative reflections of (some) participants represented in each research site, I aimed to make space for reflection between the time the project took place and the time this book would be published. I also found it important to document how participants reacted to the way our shared work is represented in this project.

The projects I introduce in this book resulted in hundreds of hours of video and audio data, digital and written artifacts ranging from definitions and translations on sticky notes to hundreds of pages of written reflections to websites collaboratively designed for each research site, as well as thousands of images of each interaction at each research space. Within each project there are multiple research goals, objectives, and desires expressed by multiple stakeholders, including the participants themselves. Thus, each data point was analyzed differently depending on the project, the research team, and their individual goals. For the purposes of my analysis in working to understand how multilingual experiences were designed and sustained across spaces, I looked at the data holistically throughout the four years total in which I was working on all three projects described in chapters 4–6. That is, while each research site had individual projects and goals embedded within it, my general research goal in these projects was to understand and trace how multilingual experiences were designed, built, and sustained for and with participants. Thus, throughout the course of the four years, I used field notes; tracked visual, written, and digital artifacts; and engaged in conversations with my participants about the role language was playing in our interactions. Much like the usual contexts in which professional translation and interpretation happen—doctors' visits, legal proceedings, parent-teacher conferences—the goal of the projects described in this book was not limited to studying language transformation alone. I wasn't focused on how each person translated or didn't translate information. Instead, through four years of observation and collaborative analysis, I sought to understand the general experience of researching in different languages, contexts, and places for different purposes through the lens of language access. Rather than engage in systematic coding of a limited data set, I engaged in participatory data analysis by tracing multilingual experiences across contexts focused on issues of language, power, and positionality—discussing these experiences with participants both during and after each project was completed and then drafting and getting feedback from participants on how their multilingual experiences are presented in this book.

In a previous project (Gonzales 2018), I identified the concept of "translation moments" as a helpful way to study language fluidity. Translation moments can be defined as instances when multilinguals pause to make a rhetorical decision about how to translate specific information for a specific audience in a specific moment in time. In that project, I counted and tracked translation moments to understand how multilingual communicators made rhetorical decisions about which words to use to make information accessible to their audiences in specific moments. While I find the notion of translation moments to be useful in understanding isolated instances of translation, in the projects I outline in this book, I aim to contextualize translation moments not only in the language transformation that took place in a single instance but also in the historical contexts that underpin the lived experiences of each group. Thus, what I present in each project chapter (chapters 4–6) are localized examples of language transformation that are also historically grounded and contextualized. To decide on which examples to include in this book, I used my field notes, observations, and visual, alphabetic, and digital artifacts to engage in conversations with various participants at each research site so we could collaboratively decide on the examples that best represent the multilingual experiences we built together. Sometimes, these conversations took place during the projects themselves; at other times, the conversations spanned many years as the projects continued and as my collaborators and participants and I shifted locations, spaces, and orientations to working together.

In discussing "the development and sustenance of relationships between researchers and participants," Timothy San Pedro and Valerie Kinloch (2017) "vehemently refuse to hide behind the façade often attempted in empirical research—that is, to seek an objective truth, our stories must have no bearing on how we come to know with others" (374). In their work in the field of education, San Pedro and Kinloch argue that researchers "must center and sustain relationships in our work. This requires that we envision research as more expansive than that which is taken up by specially trained researchers whose primary purpose is to measure validity in terms of an imagined neutrality" (374). In developing my own process of analysis for the projects presented in this book, rather than focus on specific data points, I focused on the relationships this book encompasses and on the moments and experiences that shaped and continue to shape these relationships. This means that throughout the course of four years, I kept track of meaningful experiences, stories, and examples in collaboration with several groups of people who were generous enough to welcome me into their spaces and to work with me on projects that hold meaning for many more individuals than those few introduced in the pages of this project. Emphasizing the fact that language, in its many, constantly shifting forms, shapes all human interaction and relationality, I orient to the stories presented in this book with the goals of helping other technical communication researchers recognize (1) how our presence in communities inherently shifts and moves the work a community undertakes and (2) the responsibility that engaging in community work encompasses, particularly when we are working across languages. In chapter 4, I begin by tracing multilingual experiences on the Mexico/US border alongside families and communities that have always communicated beyond and outside standardized notions of independent, static, standardized languages.

# 4
# LANGUAGE FLUIDITY IN HEALTH CONTEXTS ON THE MEXICO/US BORDERLAND

As explained in chapter 3, multilingualism is often described in binaries: this language *or* that language, Spanish *or* English, source language *or* target language. Many professional translation and interpretation protocols also operate under the assumption that to provide language accessibility, translators and interpreters should transform information from one discreet language (e.g., Spanish) to another (e.g., English). Yet as described in chapter 2, decades of research in education, linguistics, anthropology, and related fields demonstrate that languages do not function as discreet, stable categories that can be neatly segmented. Instead, languages are dynamic, fluid, and constantly changing; and individuals who speak multiple languages regularly blend and move through their entire linguistic repertoire when communicating with different audiences.

One of the primary premises of this book is that as technical communication researchers continue to engage with multilingual audiences in global contexts, researchers in the field should embrace more flexible and nuanced understandings of language that can account for wider variance and linguistic richness. Rather than thinking of multilingual research in linguistic binaries, we should design multilingual experiences that can account for a multiplicity of language and languaging (García 2009; García and Li Wei 2015) practices that reflect the communicative practices of the communities we live, teach, and work in. That is, to design accessible multilingual experiences in contemporary contexts, I argue that technical communicators should work to develop practices that allow for language fluidity, ambiguity, and even discomfort among English-dominant researchers—where the goal is not, as technical communicators may assume, to clearly transform all information into a single language for a single audience but rather to leverage multiple linguistic practices to establish trust, collaboration, and participation with multilingual communities who can guide effective multilingual experiences that lead to effective communication and social justice.

https://doi.org/10.7330/9781646422760.c004

While working toward fluid language models in technical communication is a necessary goal, establishing fluid language practices in both research methodologies and technology design is a complicated process. Both societal infrastructures and technological interfaces embrace binary approaches to language practice, positioning people, keyboard characters, writing systems, and interfaces as "either" in one language and "or" in another (Pérez-Quiñonez and Carr Salas 2021). For example, when looking at multilingual websites, users can frequently only choose to be on "the Spanish side" and then perhaps click over onto the "English side," the "Quechua side," and so on. Computer code is built on binary infrastructures that instinctively reduce communication to binaries, which means that many of the technologies currently in use, and the people who design them, sometimes also embrace this limiting approach (Pérez-Quiñonez and Carr Salas 2021). Yet as Manuel Pérez-Quiñonez and Consuelo Carr Salas (2021) argue, "We need to design interfaces to support bilingual interactions and bilingual users differently than just any user with two modalities of use" (66).

With the goal of moving toward more fluid language models in technical communication research, the primary questions I seek to answer in this chapter are: (1) how do we as technical communicators facilitate the design of tools, technologies, and information that move beyond the binary approach to language; (2) how do we use methods such as participatory design and user experience research to work with communities to design these fluid multilingual technologies that can enrich and inform effective technical communication practice in our contemporary globalized world; and (3) what does multilingual technical communication entail not within standardized categories like the singular "Spanish" and the singular "English" but across these categories in localized linguistic practices? To begin to answer these questions, in this chapter I describe what I learned from and with several community organizations and partners in the borderland city of El Paso, Texas.

El Paso borders Ciudad Juárez, Chihuahua, Mexico, and Las Cruces, New Mexico, USA, and residents have strong ties, often involving daily commutes across this metroplex area (Durá, Gonzales, and Solis 2019). As Lucía Durá, Laura Gonzales, and Guillermina Solis (2019) explain, the "total metroplex population [of this area] is estimated at 2.7 million" (1). In El Paso, "while 30% of the population speaks only English, 67% speaks Spanish or Spanish and English" (Durá, Gonzales, and Solis 2019, 1). As such, El Paso encompasses a binational, largely bilingual community with extensive expertise in fluid language practice (Nuñez 2019).

To begin to illustrate the design of multilingual technical communication experiences on the Mexico/US border, I'll provide historical context on this borderland region, focusing specifically on how language relations developed and continue to emerge in this binational community. I ground this discussion in two health-related projects I co-conducted in this context, including a collaboration with a public health grant–funded project related to diabetes treatment and prevention and an after-school program for youths and families focused on health and wellness literacy. Through these examples, I illustrate multilingual technical communication experiences on the border as those that require (1) a reliance on and centering of community expertise, (2) the establishment of trust and relationality among designers and community members, and (3) an emphasis on fluidity rather than correctness in design.

## COMMUNITY RESEARCH SITES AND POSITIONALITY

The two research sites I introduce in this chapter represent communities of people I had the privilege to work with as part of a larger collaborative project focused on localizing healthcare on the Mexico/US borderland through a multilingual user experience design (Durá, Gonzales, and Solis 2019). Along with my collaborator Lucía Durá, we developed a multi-site study geared toward developing linguistically and culturally sustainable tools and technologies for facilitating health-related communication and interventions for bilingual populations in the borderland city of El Paso. Decades of research in health communication and more recently in the subfield of the Rhetoric of Health and Medicine (RHM) demonstrates that more localized approaches are needed to provide accessible care for bilingual and multilingual communities (Bloom-Pojar 2018; Cristancho et al. 2008; Durá 2016; Gonzales and Bloom-Pojar 2018). As Sergio Cristancho and colleagues (2008) explain, "For Hispanics living in the United States, barriers such as lack of health insurance, non-eligibility for certain public assistance programs, costs of health care services, language, lack of medical interpretation services, discrimination related to documentation status, and lack of transportation" are well-documented barriers to access to and use of healthcare (634). These barriers are exacerbated for borderland residents who experience discrimination and xenophobia in an increasingly policed state and whose language practices do not fall neatly into either formal English or formal Spanish categories (Durá, Gonzales, and Solis 2019; Concha 2018; Gonzales and González

Ybarra 2020). While state-sanctioned healthcare tools and information in the borderland are often ineffective, borderland residents themselves have extensive expertise in fluid approaches to language that can inform localized design of healthcare interventions (Durá, Gonzales, and Solis 2019). Recognizing the potential for local, community-driven innovation in the borderland region, Lucía and I, along with multiple other collaborators, set out to work with borderland residents to develop localized tools for facilitating healthcare in multiple areas through the practices participatory design and user experience.

All of the collaborators on this project identified as Spanish-English bilinguals, with several collaborators (including Lucía) identifying as Mexican and/or Mexican American or Chicanx. As a bilingual Bolivian immigrant, I shared some linguistic expertise with participants in these projects. Yet I very much credit this borderland community for teaching me how to practice and embrace fluid communicative practice across Spanishes and Englishes.

For the purposes of this chapter, I focus on two specific projects that fall within this broader healthcare on the border initiative. The first is a project titled "the Diabetes Garage," which is a culturally tailored diabetes self-management and education/support program that integrates automotive maintenance/repair analogies to increase men's interest in diabetes programming. Led by Dr. Jeannie Concha, assistant professor of public health at the University of Texas at El Paso (UTEP), the Diabetes Garage seeks to serve Latino men in diabetes treatment/prevention programs, seeking to target this historically underserved population that is currently burdened by diabetes at a prevalence that is 150 percent higher than that of non-Hispanic whites (Concha 2018). In the fall and spring of 2018, Lucía and I and our research team partnered with Jeannie to conduct usability tests and participatory translation focus groups regarding some of the materials that were being designed for the Diabetes Garage. This includes a brochure intended to translate diabetes-related information into "car language" in Spanish. For example, the Diabetes Garage uses the metaphor "check your engine" on the brochure to motivate men to check their blood glucose levels. Our job as a research team in this collaboration was to then help translate the metaphorical "check your engine" into Spanish for and with this borderland community.

The second research site I introduce in this chapter is La Escuelita, an after-school health literacy program for youths and families that I co-facilitated from 2016 to 2019. The program, which started in 2012 through an interdisciplinary collaboration between university and housing authority partners, takes place in public housing community spaces in El Paso. For the three years I was involved as co-facilitator, the goal of La Escuelita was to build culturally localized health practices that honor participants' transnational backgrounds and heritage (see Del Hierro et al. 2019; Flores-Hutson et al. 2019). Participants in the program shifted slightly every year, as our meeting locations changed based on where the participating families were moved to by the local housing authority. While there was some fluctuation in participants, for the three years in which I was directly involved in the program, participants consisted of five to eight transnational families who moved regularly between their homes in El Paso and Ciudad Juárez. For example, some of our participants were youths and their mothers who lived in El Paso during the school/work week and then commuted to Juárez to spend time with their fathers/husbands on the weekends. Due to issues of documentation, finances, and more, many of our participants' families and loved ones resided on either side of the border, while our participants themselves commuted back and forth from El Paso to Ciudad Juárez when they could.

La Escuelita met once per week in a community center located in our participants' neighborhood, which was one mile from a port of entry into El Paso, Texas. The community center was in the Chamizal neighborhood, named after the national parks that sit on each side of the Mexico/US border, separated by the Rio Bravo/Rio Grande (Del Hierro et al. 2019). During our weekly meetings, program leaders (including me and other faculty members from my academic institution) facilitated various activities that interrogate notions of health, wellness, and nutrition through mediums such as art, community-based mapping, and collaborative cooking. All of these activities took place in both Spanish and English, as all of our participants identified as bilingual. While I report with my collaborators on other findings and elements of this project in other venues (see Del Hierro et al. 2019; Flores-Hutson et al. 2019), in this chapter, I focus specifically on the translation, technical communication, and user experience elements encompassed in this project, which allowed me to trace and develop strategies for designing multilingual technical communication experiences in this borderland context.

While the Diabetes Garage and La Escuelita are two individual projects with very different participants, the multilingual experiences created and sustained in both spaces provide helpful models for how technical communication researchers can continue to move beyond binary notions of languages in our research and design practices. To understand how multilingual experiences are designed and shaped by local communities, it's important to understand the historical underpinnings of how language is used in current contexts. As such, in the sections that follow, I share examples from the Diabetes Garage and La Escuelita in the context of both historical and contemporary language relations on the border.

## LANGUAGE RELATIONS ON THE MEXICO/US BORDERLAND

As many borderland scholars demonstrate, while residents of the Mexico/US border engage in fluid language practice across varieties of Englishes and Spanishes on a regular basis, these fluid linguistic movements represent rhetorical choices that can stem from long histories of colonization and racial violence (Anzaldúa 1987; Mignolo 2000). As Gloria Anzaldúa (1987) explains, borderland identity is much more than just a physical boundary across nations; "a borderland is a vague and undetermined place created by the emotional residue of an unnatural boundary" (15). As such, as Gonzales and Mónica González Ybarra (2020) elaborate, "the borderlands are in-between, highly contested spaces in which marginalized peoples (Black, Indigenous, people of color, women, queer folks, and youth) navigate, survive, and cultivate oppositional consciousness (Sandoval 2013) to resist oppressive structures (de los Ríos 2018; Nuñez 2019)" (228). At the same time, as Ariana Brown (2021) illustrates, conversations and discussions of borderland identity often erase Black Mexican experiences, ignoring the anti-Black racism that extends across both sides of the Mexico/US border.

Thus, one of the first things to know about studying language on the border is that the fluidity among different varieties of Spanishes, Englishes, and other Indigenous languages is not always practiced as an adornment to self-expression or as a performance of identity. Instead, language fluidity, particularly for racialized borderland residents who experience racism and xenophobia from both the US and Mexico, is rhetorical action grounded in lived experience. As Gonzales and González Ybarra (2020) explain, drawing on Walter Mignolo's (2000) concept of "border thinking," language practices on the Mexico/US borderland, and in El Paso specifically, can offer "a reframing of knowledge production that centers those living in and on the margins—those who have knowledge and experience resisting and navigating colonial dominance" (228). For some, to translanguage, or move fluidly across Spanishes and Englishes, in this borderland context can be an act of resistance to standardized language ideologies imposed on borderland residents while also being the consequence of long-standing discriminatory practices against Mexican Americans in both the US and Mexico (Nuñez 2019).

According to Oscar Jáquez Martínez (1994), "A border is a line that separates one nation from another or, in the case of internal entities, one province or locality from another. The essential functions of a border are to keep people in their own space and to prevent, control, or regulate interactions among them. A borderland is a region that lies adjacent to a border" (5). As Margarita Hidalgo (1986) clarifies, El Paso and Ciudad Juárez are "two border cities [that] share a common life that dates back to the seventeenth century, when the two communities, known as Paso del Norte, were one. Although the Mexican-American War (1846–48) separated the two communities politically, they have remained closely linked by social, economic, and cultural forces" (194). This "1900-mile Mexican-US boundary" is a complex and interesting site to study language and cultural fluidity, "a manifold language setting which involves the use of two official languages—Spanish and English, their regional and social varieties, and more interestingly, the blending of the two in the vernacular" (Hidalgo 1986, 193).

From a sociolinguistics perspective, language on the Mexico/US borderland has been studied in "four areas: (1) code switching between English and Spanish within the context of a conversation; (2) typologies and characteristics of Spanish, English, and intermediate codes in use in the community; (3) the use of special Chicano linguistic creations, such as pocho, calo, etc., and (4) the child's acquisition of Spanish and English" (Peñalosa 1980, 6). As early as 1980, sociolinguist Fernando Peñalosa (1980) clarified that "such studies have established at least that Chicano linguistic behavior is neither erratic nor the product of confused minds torn by culture and conflict, nor do Chicanos suffer from congenital deficiencies of cognition and conceptualization [as was and sometimes still is widely assumed], despite widespread stereotypes found in the social science and educational literature" (6). Yet despite contrasting linguistic evidence, the languaging (García and Li Wei 2015) practices of Chicanx communities have been historically stigmatized, with stigma and marginalization increasing in the US political climate under the Trump presidency and the xenophobic raciolinguistic ideologies of an administration that positioned Mexican people, cultures, and languages as inferior to white, Western American, standardized English-dominant practices. As Cati V. de los Ríos (2018) argues, "A sociopolitical climate of hostility . . . only intensifies the urgency to study language and literacy practices as forms of resistance, transformation, and possible solutions" (457). Further, the languaging practices of Chicanx communities are marginalized in the design of tools and technologies for bilingual Spanish/English speakers, designs that have historically been localized for speakers of either formal, standardized Spanish or formal, standardized English (Nuñez 2019).

Many researchers have analyzed how perceptions of Spanish and English on the Mexico/US border are influenced by the broader cultural and ideological histories and experiences of communities that live between two nations (Hidalgo 1986). For example, Patricia MacGregor-Mendoza (1999) examined how English-only policies embraced in school settings resulted in the loss of the Spanish language for many US residents of Mexican descent. Linguists such as Roberto A. Galván and Richard V. Teschner (1977) conducted longitudinal studies on "border language," developing the *Dictionary of Chicano Spanish*, which includes over 9,000 Spanish terms and idioms developed in the borderland region. As Peñalosa (1980) explains, "The study of the Chicano's use of language, in both historical background and social context, has long been a concern of scholars, practitioners, and many average citizens" and could, as I demonstrate in this chapter, be then adapted to influence the successful design and localization of borderland health tools and technologies (1). This type of application and incorporation of borderland language practices into contemporary technology design requires a complete re-shifting and restructuring of established infrastructures for information design that currently follow colonial, standardized linguistic practices.

While the fluidity, power, and constant evolution of borderland language have been studied and recognized extensively in academic venues and popular media, the everyday communicative realities and perceptions of language by people who live on the border are influenced by long-standing, racialized prejudices against Mexican communities in the United States. As Hidalgo (1986) explains, "Contact serves as an effective catalyst in bringing out the beliefs, values, prejudices, and contradictions of a speech community" (193). As communities that are often perceived as neither "fully" American (i.e., stemming directly from English monolingual culture) or "fully" Mexican (i.e., monolingual Spanish speakers), Chicanx communities develop and enact hybrid language practices that echo the constant interaction between English and Spanish (and multiple Indigenous languages) that was developed through colonization of Mexican territory and has continued to evolve on the borderlands through colonization, trade, business, tourism, and multiple other factors.

Chicanx communities can be defined as "persons of Mexican descent who are resident in the United States" (Peñalosa 1980, 2). However, this notion of residency and nationality becomes complicated on the border, particularly in places like El Paso, Texas, and Ciudad Juárez, Chihuahua, where it has been largely documented that many people (professionals, students, and general residents) move fluidly between and reside on either side of the border depending on the day of the week, the time of year, their particular work, school, and personal obligations, and the political climate. Thus, to define Chicanxs by residency limits the identities of borderland community members, in the same way that describing Chicanx communities as speakers of English, Spanish, or bilingual speakers of both English and Spanish would also be limiting. In *We Are Owed*, Ariana Brown (2021) further highlights the complexity of Chicanx identities, pointing out how labels like "Chicanx" also erase Black people's existence and experiences on the Mexico/US border. Indeed, as Peñalosa (1980) explains, "The unyielding structure of social institutions has yet to introduce notions of sociolinguistic relativity into its rigidly dichotomized concept of English-Spanish alternatives" (6). In other words, social institutions (and their accompanying documents) often account for language practices that are either "English" or "Spanish" or perhaps bilingual (English *and* Spanish). However, these institutions (and I argue the language through which we define contemporary communicative practices) are not yet accustomed to or capable of understanding or identifying language beyond these fixed categories (what García and Li Wei [2015] may define as "named languages" such as Spanish, English, French, and others).

Due to the constant contact and movement between different varieties of English and Spanish experienced through life on the border, borderland residents (who de la Piedra, Araujo, and Esquinca [2018] define as "transfronterizxs") sometimes develop particular languaging practices that draw on Spanish, English, and multiple Indigenous languages. These languaging practices have been more recently defined by educational linguists such as Ofelia García (2009) through concepts like translanguaging, which is the "act performed by bilinguals of accessing different linguistic features or various modes of what are described as autonomous languages, in order to maximize communicative potential" (140). Thus, through translanguaging, borderland communities use, deploy, modify, and alter standardized Spanish and English to communicate effectively in specific rhetorical contexts in specific moments (Gonzales 2018). In the sections that follow, I introduce examples of translanguaging in the El Paso–Ciudad Juárez borderland as they were enacted by borderland residents from various demographics (in terms of gender, age, and linguistic expertise) who participated in technical communication and technology design projects related to health on the border. Through these examples, I make recommendations for designing multilingual technical communication experiences that transcend standardized language categories (e.g., standardized Spanish, standardized English) to reflect fluid and emergent community expertise that draws on multilingual expertise. Embracing an intersectional, interdependent methodology to study language on the border can provide a critical lens through which researchers can understand the connections among race, power, language, and access in borderland contexts.

## USING "PROPER" SPANISH

Participant 1: "And in the old days they used to say like if you respected, you talk to your older people, like 'de usted.' And nowadays, they talk, especially people who are bilingual, who got the bilingual thing, they actually use 'tu' instead of 'usted.' It's kind of like old-fashioned, I think."

Participant 2: "Well, I mean if you use the 'usted,' that is the proper, you know, Spanish. But if you use the 'tu' [and] um [words like] 'checa' instead of 'revise,' you know, you're . . . I think it should be more formal, I mean 'cause you are not, it is just not the people here in El Paso that you are trying to reach . . . it's the people that are actually coming from, you know, the Mexican side that are actually already here that are, that were raised with the proper Spanish."

As part of the Diabetes Garage project, I co-facilitated participatory translation focus groups with Latino men in El Paso. The purpose of these focus groups was to collaboratively translate into Spanish the car-related metaphors developed through the diabetes treatment and prevention program. This terminology was to be distributed in a brochure about the Diabetes Garage to motivate bilingual Latino men in El Paso to want to enroll in the program and learn how to manage or prevent diabetes through an engaging program tailored around cars. The two opening quotes are excerpts from a participatory translation focus group conducted with self-identified bilingual Latino (Mexican and Mexican American or Chicanx) men in El Paso who volunteered to help translate the Diabetes Garage brochure. These participants were asked to provide feedback regarding the design and translation of the bilingual brochure.

In this particular conversation, focus group participants were discussing the "type" of language the brochure should include. Although participants were given some background on the project and told that they themselves were representative target users (bilingual Latino men of varying ages who live in El Paso, Texas) for this brochure and for the broader program, as evidenced in the quoted excerpts, men engaged in a discussion about the use of "formal" versus "informal" terminology, referencing various levels of formality and propriety and their connections to Chicanx culture. The discussion of "formal" (standardized) and "informal" (fluid/borderland) Spanish was prevalent throughout all of the participatory translation focus groups conducted for this project. For example, participants mentioned that formal Spanish would cause the diabetes-related information to be taken more seriously or "with respect." At the same time, however, some participants mentioned that using less formal language, such as the word "*checa*" in "*checa tus niveles*," would make the brochure more accessible to the El Paso audience. As one participant mentioned, "You don't want to gear this [brochure] to people with master's degrees" but should focus instead on reaching broader audiences.

On the surface, these excerpts illustrate participants' helpful contributions and the thoroughness of their user feedback; these men were concerned with designing a brochure that would be appealing and usable in their community, as they, too, recognized the need for more diabetes-related interventions and programs in the area. Yet these excerpts and the broader conversation also point to ongoing questions, issues, and consequences related to language fluidity, racial relations, colonialism, and diversity in this borderland region (de la Piedra, Araujo, and Esquinca 2018).

For example, during this particular conversation, participants discussed possible translations of the word *check* in reference to the notion of "checking your blood glucose levels" within diabetes treatment and prevention. At first, the men (all of whom identified as bilingual Spanish and English speakers) suggested the word *checa* as a colloquial term frequently used to reference the English term *check* in this borderland region. Although the Castilian-derived translation of "check," according to the real academia Española on which most "formal" translations are based, may be something like the word *revisa* (closer to the English term *review*), focus group participants initially suggested the term *checa* as a colloquialism that would appeal to and resonate with local users of the brochure. Participants mentioned that they use the word *checa* in their homes in reference to "checking" anything ranging from their car engines to their bank accounts. However, later in the conversation, participants such as participant 2 quoted above questioned the suggestion to use the term *checa* in the proposed brochure, wondering if the colloquial *checa* would be offensive to Spanish speakers who do not speak the "*pocho*" Spanish (Medina 2014) found in El Paso. As participant 2 elaborated, "The people who are coming from, you know, the Mexican side" are the individuals who speak "proper Spanish" and are thus those who may be offended or put off by the colloquial term *checa*. Another participant mentioned that the brochure should contain "the proper, proper Spanish" rather than the Spanish used by "people like me or my dad" (referring to El Paso residents who speak "pocho" Spanish).

This brief but important conversation among focus group participants references several issues related to language, race, culture, and access on the Mexico/US borderland. These issues include the relationships between "the Mexican side," as participant 2 explains, or individuals who live in or were predominantly raised in Ciudad Juárez, Chihuahua, Mexico, and those who live in and/or were predominantly raised on "the American side," in El Paso, Texas (Hidalgo 1986). A long history of research highlights how borderland residents, and Chicanx communities specifically, are frequently positioned as not "from" Mexico or from the US and therefore as not speaking "proper" Spanish or "proper" English. Although the purpose of the Diabetes Garage was to target material specifically for the Chicano men represented in the participatory translation focus groups, language relations in this region, and in the US and Mexico more broadly, consistently degrade non-standardized Spanishes in favor of standardized variations rooted in the European Castilian. This push to favor standardized language practices also extends to the design of technologies and to the development of digital algorithms that shape information searching and retrieval. For example, if a borderland resident wanted to Google how to "check" their glucose levels, they would have to use the formal Spanish word *revisar* rather than the borderland term *checar* to get more accurate results. As Safiya Umoja Noble (2018) makes clear in her discussion of *Algorithms of Oppression*, search engines like Google, in both their Spanish and English versions, are built on oppressive infrastructures that favor white/Western/standard language ideologies. Searching for the term *checa* on Google yields results that are unrelated to the meaning of the word *check*. Meanwhile, searching for the word *revisa* on Google will yield accurate definitions of this term.

Borderland residents like those involved in the Diabetes Garage participatory translation focus groups are already in the practice of retrofitting their own language preferences when working with technological interfaces to find information, meaning they will adjust their preferences when searching for information in formalized spaces. While many individuals can and do adjust to searching for information in "formal" Spanish, the lack of availability of localized language tools and health-related information contributes to ongoing health disparities imposed on borderland residents. For the participants in the Diabetes Garage focus groups, there was a stark difference between "formal" and "informal" Spanish, and health-/science-related information needed to be presented in standardized Spanish to be credible to Spanish-speaking audiences in the borderland region. As evidenced in this discussion, some members of this borderland community already recognize that technical information will not reflect their own linguistic practices, as the raciolinguistic ideologies embedded in these tools will always favor colonial approaches to language historically and contemporarily deemed superior (Bloom-Pojar 2018). As the Diabetes Garage participants explained, if people in El Paso want accurate health-related information, they will likely have to use the "formal, formal Spanish" to search for information and will thus need to be able to read and understand information in standardized Spanish to seek adequate healthcare. The notion of language access in the context of the participatory translation focus groups at the Diabetes Garage was shaped by racialized underpinnings that both historically and contemporarily shape the suggestions borderland residents will make in collaborative technology design projects.

## BILINGUAL WEB DESIGN AT LA ESCUELITA

While one might attribute the emphasis on "proper" Spanish embraced by participants in the Diabetes Garage project to participants' ages and lived experience, the youths who participated in La Escuelita also exhibited a critical awareness of the differences between "formal" and "informal" Spanish and their role in communicating with various audiences. Part of our work with La Escuelita included building a website to highlight the activities and perspectives of this borderland community, specifically in relation to localized notions of health and wellness. Over the course of three years, other facilitators of La Escuelita and I incorporated various activities into our meetings that helped us build a website for the group, which is temporarily hosted at www.escuelitaep.com.

While many elements and moving pieces were involved in both facilitating and researching at La Escuelita, one of the main deliverables we sought to design, through funding from a research grant, was our bilingual website that represented notions of health and wellness as they were understood by borderland residents like our participants. The purpose of this website, as identified by our participants in collaboration

La Escuelita es una familia... pero además es una asociación entre los residentes de Sherman y los profesores de UTEP. La Escuelita es programada por El Paso Housing Authority. Nosotros nos juntamos una vez por semana para hablar, aprender, y para compartir actividades sobre la comida, salud, y cultura. Creamos esta página para compartir con nuestra familia todos nuestros recuerdos durante el año. También compartimos recetas, actividades, y la información que aprendemos durante el año.
La Escuelita is a family...but also a community partnership between residents of the Sherman neighborhood and UTEP Professors that is facilitated by the El Paso Housing Authority. We get together once a week to talk, learn, and share activities on food, wellness, and culture. We created this page to share with our familia all the memories from throughout the year. We also want to make sure we share all the recipes, activities, and information we learned throughout the year.
FOOD
WELLNESS
CULTURE

*Figure 4.1. La Escuelita's landing page.*

with our research team, was to depict what health and wellness means for participants at La Escuelita—all of whom seek to counter white, Western conceptions of "healthy" by situating health and wellness in culturally localized practices such as cooking traditional meals and researching Mexican herbs and superfoods (e.g., chia, spirulina), as well as practicing mindfulness and meditation through art and fitness activities (see Flores-Hutson et al. 2019).

While we engaged in many activities to build the site along with our participants, issues of translation and localization came up consistently in our discussions, particularly when we engaged in participatory translation sessions similar to those practices at the Diabetes Garage. For example, over several Escuelita meetings, our group (which consisted of bilingual youths ranging from pre-K to grade 12 as well as mothers, grandmothers, and our research team) worked to write and translate content for the Escuelita landing page, which is depicted in figure 4.1.

As evidenced in the screenshot depicted in figure 4.1, Spanish and English are used consistently throughout the Escuelita website. At times, Spanish and English translations are provided side by side on the site, with Spanish always listed first. Other times, however, content is provided only in Spanish or only in English. Unlike many traditional bilingual websites, La Escuelita's website does not contain two portions, one in English and one in Spanish. Instead, different variations of Englishes and Spanishes are used throughout, based on the community's rhetorical translation decisions.

The text on our landing page was originally written in English through conversations with all members of our group. The text in English reads: "La Escuelita is a family . . . but also a community partnership between residents of the Sherman neighborhood and UTEP professors that is facilitated by the El Paso Housing Authority. We get together once a week to talk, learn, and share activities on food, wellness, and culture. We created this page to share with our familia all the memories from throughout the year. We also want to make sure we share all the recipes, activities, and information we learned throughout the year."

Participants at La Escuelita felt the initial text on our website should be largely in English, with the exception of the word *familia* that represents the group's relationship as family. Although, according to our participants, English should be listed on the site first so that "more people can understand the information when they visit the website," participants also felt the text should be translated into Spanish. The Spanish content, which was later placed above the English content based on our participants' suggestions, reads: "La Escuelita es una familia . . . pero además es una asociación entre los residentes de Sherman y los profesores de UTEP. La Escuelita es programada por El Paso Housing Authority. Nosotros nos juntamos una vez por semana para hablar, aprender, y para compartir actividades sobre la comida, salud, y cultura. Creamos esta página para compartir con nuestra familia todos nuestros recuerdos durante el año. También compartimos recetas, actividades, y la información que aprendemos durante el año."

Our Escuelita collaborators felt we should provide a Spanish translation for visitors who come to the site and want to learn more about what we do. To translate this one initial welcoming paragraph, members of La Escuelita engaged in multiple participatory translation sessions, where we all discussed potential translation options and made decisions about the content we should include for our community. For examples, figures 4.2, 4.3, 4.4, and 4.5 are images from one of our participatory translation sessions, where each member of La Escuelita had a printout of our landing page that displayed only the English content. Each participant then translated the content into Spanish on their own sheets of paper before we collectively discussed the translations as a group.

The handwritten translations presented in figures 4.2–4.5 all include some differences in translation choices, and they were completed by participants that included youths (from middle school through high

*Figure 4.2. A mother at La Escuelita translates our website landing page.*

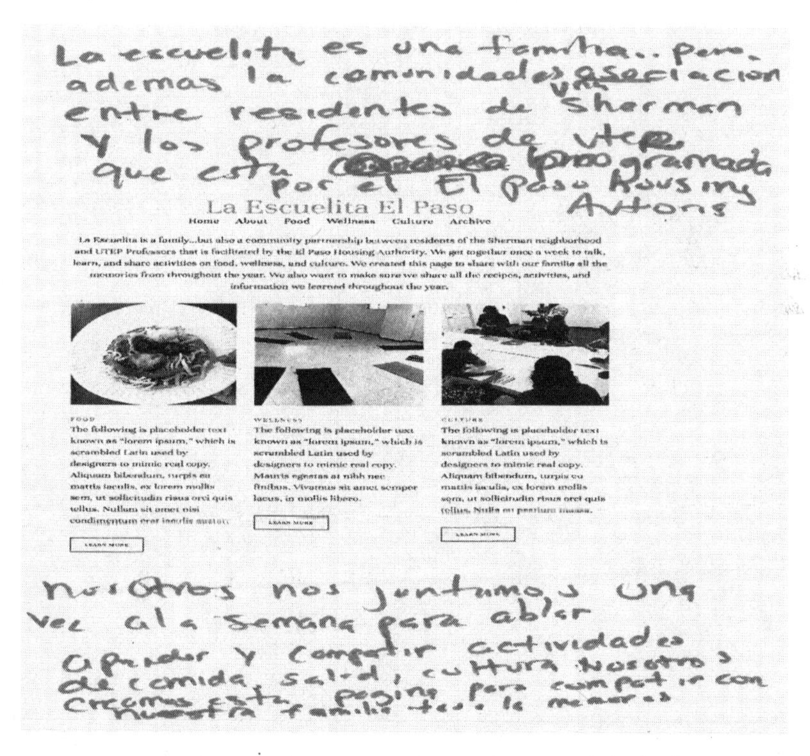
La Escuelita El Paso
Home About Food Wellness Culture Archive
La Escuelita is a family...but also a community partnership between residents of the Sherman neighborhood and UTEP Professors that is facilitated by the El Paso Housing Authority. We get together once a week to talk, learn, and share activities on food, wellness, and culture. We created this page to share with our familia all the memories from throughout the year. We also want to make sure we share all the recipes, activities, and information we learned throughout the year.
FOOD
WELLNESS
CULTURE
LEARN MORE

*Figures 4.3, 4.4, and 4.5. Youths at La Escuelita translate our website landing page.*

La escuelita es unafamilia... pero tambien una comunidad de compañerismo entre los residente de los apartamentos Sherman y los profesores de UTEP que es proveida por El paso Housing authority". Nos reunimos cada semana nos reunimos cada semana y platicamos, aprendemos, y compartimos actividades con comida, salud, y cultura. Creamos una Pagina para compartir con nuestras familias las memorias de todo el año. tambien compartimos las recetas y actividades y informacion que aprendimos todo el año

La Escuelita El Paso

Home About Food Wellness Culture Archive

La Escuelita is a family...but also a community partnership between residents of the Sherman neighborhood and UTEP Professors that is facilitated by the El Paso Housing Authority. We get together once a week to talk, learn, and share activities on food, wellness, and culture. We created this page to share with our familia all the memories from throughout the year. We also want to make sure we share all the recipes, activities, and information we learned throughout the year.

FOOD
The following is placeholder text known as "lorem ipsum," which is scrambled Latin used by

WELLNESS
The following is placeholder text known as "lorem ipsum," which is scrambled Latin used by

CULTURE
The following is placeholder text known as "lorem ipsum," which is scrambled Latin used by

La Escuelita es una familia... pero tambien una communidad colvaracion entre familias en el Sherman de vesinos y Utep profesores que esta ecordinada de El Paso Housing de deredidiensas. Nos juntamos una ves a la semana para ablar, aprender y compartir actividades de comida, salud, y cultura. Nosotros creamos esta pajina para compartir todas nuestras resetas, actividades y information nosotros aprendemos durante todo el año.

La Escuelita El Paso

Home About Food Wellness Culture Archive

La Escuelita is a family...but also a community partnership between residents of the Sherman neighborhood and UTEP Professors that is facilitated by the El Paso Housing Authority. We get together once a week to talk, learn, and share activities on food, wellness, and culture. We created this page to share with our familia all the memories from throughout the year. We also want to make sure we share all the recipes, activities, and information we learned throughout the year.

FOOD
The following is placeholder text known as "lorem ipsum," which is scrambled Latin used by

WELLNESS
The following is placeholder text known as "lorem ipsum," which is scrambled Latin used by

CULTURE
The following is placeholder text known as "lorem ipsum," which is scrambled Latin used by

school) and their mothers. For example, the initial phrase "La Escuelita is a family . . . but also a community partnership" was translated by different participants as follows: (1) "La Escuelita es una familia . . . pero ademas la comunidad es una asociacion," (2) "La Escuelita es una familia . . . pero tambien una comunidad de compromiso," and (3) "La Escuelita es una familia . . . pero tambien una comundad colvaracion entre familias." Each of these translations includes what may be perceived as "errors" in grammar or style, but what they really represent are three different approaches to understanding the relationship between participants at La Escuelita and the organizations that support this project. Each translation was completed by participants of different ages with different linguistic and translation histories, and those histories are reflected in the written translations and were also valuable in our participatory translation discussions.

Although all participants at La Escuelita identify as bilingual speakers of Spanish and English, each member of La Escuelita has a different language history and feels different strengths communicating in Spanish and/or English. During the participatory translation focus activities, we engaged in discussions about our different translation choices and how these choices reflected how we felt about La Escuelita and how we used words to express ourselves. Each member of the group shared their translation with the group and explained their particular translation choices, and the group engaged in a conversation about the various translation options presented by the team and how the team should decide which version(s) of the translation to include on the final website. Practicing translation in community helped the group shape the identity of La Escuelita in a public-facing website while also helping us articulate to each other what La Escuelita meant to each participant. Further, practicing collaborative translation, as one mother explained during this activity, "help[s] us see that we have skills that others don't," meaning that translation was a way to demonstrate language skills that are often dismissed in English-dominant contexts. Furthermore, participatory translation activities allowed both youths and mothers to contribute to the discussion, noting the interdependent nature of accessible bilingual web design. Since mothers and youths had different experiences with Spanishes and Englishes, each group could make contributions geared toward different potential users of the Escuelita site. It was only through collective conversations about these translation suggestions that both youths and mothers were able to decide on which translations would be most effective in specific portions of the website.

On most of the Escuelita website, participants express their different languaging practices by including content in variations of Spanishes and Englishes. However, as we discussed the design of the site, participants explained that the landing page should be "the most formal" section of the website because, as one middle school–age participant explained, "this is the first thing people will see about us." As a result, the landing page is one of the only areas on the site where language is the closest to standardized Spanish and English. In our participatory translation activities, participants spent time translating both individually and collaboratively, eventually deciding, through multiple conversations and translation share-alouds, on the translation we currently display on the site (depicted in figure 4.2). Although the language on this landing page is relatively standardized, the language we chose to display on the site still reflects the communicative conventions of a borderland community that includes members of various ages. Through their participatory translation activities, youths and their mothers and grandmothers at La Escuelita demonstrated an awareness of the language practices they could use in our program community versus the language practices outsiders would expect on a formal website. Like participants in the Diabetes Garage who made language suggestions using colloquial terminology such as "checa" within the group of participants, La Escuelita participants also created translations that reflected their own fluid borderland practices. For both sets of participants, however, placing their translations on public-facing platforms like a website or a brochure for public distribution meant the fluid borderland language practices should be (at least partially) replaced by standardized Spanish and English norms—a linguistic choice that echoes what borderland communities have been hearing and experiencing in formal education and communication spaces that function through a standardized, binary, and white/Western linguistic framework.

## TRANSLATING WITH VISUALS TO MOVE BEYOND LANGUAGE BOUNDARIES

While participants in both the Diabetes Garage and La Escuelita discussed the differences between "formal" and "informal" Spanishes in written form, both groups also seemed to embrace more fluid approaches to language when incorporating visuals into their designs. As Bloom-Pojar and Danielle DeVasto (2019) explain, "Visuals are often used to supplement or replace written or verbal language for quick, accessible communication. This perspective has been especially true in cross-cultural healthcare situations, where visuals have been endorsed as strategies for addressing the challenges and complexities of communicating with audiences who may have varied ways of speaking and reading texts" (n.p.).

In both the Diabetes Garage and La Escuelita, participants used visuals to expand rigid conceptions of formal, standardized, alphabetic communication design. For example, during the Diabetes Garage participatory translation focus groups, participants spent extended periods of time commenting on the visuals included in the brochure, even though the focus of the questions was predominantly on the linguistic translations of the content. The brochure depicted the human body in the shape of a car dashboard, with labels such as "check your [blood glucose] levels on the speedometer" and "check your fuel" (meaning food intake versus activity level) on the odometer. Participants zoomed in on these visuals and made re-design suggestions, making comments such as "this [visual] is kind of confusing for somebody who might not recognize the gauge," "I guess they [the designers] are trying to go with the odometer there, but . . . it is not coming through," and "I find confusing the placement of the hands of the gauge, because if people don't necessarily read the instructions or directions, they might be thinking like having a blood glucose of 300+ is okay or something." Frequently during these discussions, participants would get up from their seats and use a white board to draw or sketch out their design suggestions, emphasizing the importance of accurately representing health-related information in visual form.

Throughout the discussions, participants emphasized that users of this brochure would be focusing on the images just as much as, if not more than, the words. Participants mentioned that if the designers of the brochure wanted to "really make it [the human body metaphor] look like a car, then play around with the colors to make it look real, make it look more like a dashboard with a metal finish." While the Diabetes Garage brochure was intended to look like a car dashboard, participants mentioned that the gray color used in the design was not realistic, as it needed to be silver and shiny to truly communicate that "the body should be cared for and remain healthy and shiny like our cars."

Other participants stated that the design of the brochure should be emphasized more and made more realistic "for somebody to get the information." As one participant elaborated, "I think a lot of people, if you just give them this, they [will] say 'ah, that is a lot of information.' " In other words, participants recognized that even though the translation of the written information on the brochure is important, the visual design needed to appeal to users just as much as, if not more than, the written words. As participants argued, users of the brochure would not find information accessible or interesting if the surrounding visuals are not engaging and realistic.

At La Escuelita, visuals also became an important element of our collaborative multilingual design. One of the activities we frequently practiced at La Escuelita was creating drawings or images to represent different types of exercises or activities youths and their families practiced to stay well and healthy. Sometimes, we would practice exercises like yoga together as a group and would then ask participants to draw other types of physical activity they enjoyed doing at home or in the park. For each of these images, the group worked to create captions and image descriptions that would help us ensure that the images would be accessible for everyone in the group, especially if we selected these drawings to be featured on our website. For example, during one activity, each participant was asked to draw and then describe something that makes them feel well and healthy. Valerie, a ten-year-old participant, drew the image depicted in figure 4.6, which she titled "Gimast-ick." She also drew and colored a picture of the Mexican flag to represent something that made her feel happy.

As depicted in figure 4.7, Valerie then captioned her drawings "Lo que yo dibuje es de gimnastic. Me gusta hacer pero no me se la split. I also drew the Mexico flag because it is important to me." In this brief caption, Valerie translanguaged between Spanish and English to describe the things that make her feel well, healthy, and happy—doing gymnastics and the Mexican flag. In her caption, Valerie did not provide translations for the things she wrote (in either Spanish or English). Instead, she leveraged her linguistic knowledge by incorporating both languages within a single caption, thus illustrating her binational identity in both her drawings and her writing and incorporating this identity into the description she provided to make her drawing accessible to the group.

Valerie did not translate the word *split* into Spanish because, as she mentioned, "I didn't know how to say it so I just say split." Frequently, participants at La Escuelita mentioned that they wrote down or vocalized words in the language that came to their mind first, without worrying about whether they were "supposed" to be communicating in one language or the other. In cases where participants could not think of a specific word in a specific language right away in our discussions or activities, participants would often ask the group, "y como se dice ____ en Inglés" or "and how do you say _____ in Spanish?" In these instances, members of the group would respond with possible translations, frequently leading

*Figure 4.6. Valerie draws gymnastics.*

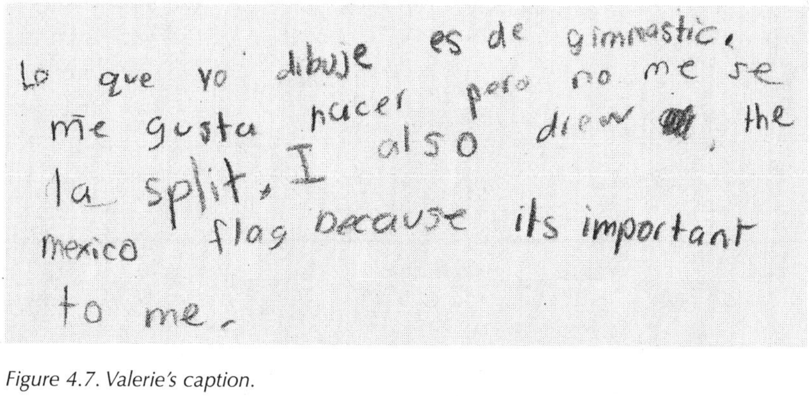

Lo que yo dibuje es de gimnastic. me gusta hacer pero no me se la split. I also drew the mexico flag because its important to me.

*Figure 4.7. Valerie's caption.*

to conversations about how a word could or should be translated for a specific purpose in a specific context. In this way, establishing accessibility in our designs at La Escuelita consisted of combining visual and written communication with the expertise of the group as a whole. Designing multilingual experiences in this context encompassed the incorporation of translation activities that allowed participants to feel comfortable expressing themselves in whatever language(s) they preferred while also feeling comfortable enough to ask for assistance or clarification when translating specific concepts or ideas for the group.

The combination of visuals and written content also became helpful for participants at La Escuelita as the group engaged in other technical communication and participatory design activities. For example, in one activity, we asked all participants to write down instructions (i.e., recipes) for making their favorite meals. The purpose of this activity was for the group to think about and discuss how food is traditional and culturally localized and how eating healthy and staying well does not necessarily mean we need to give up our cultural foods and cooking practices.

For this activity, the group used journey mapping and affinity diagramming, two common practices in user experience design, to both write and design their instructions. Participants would first trace their journeys in learning a specific recipe by writing about who taught them the recipe, where they learned the recipe, and how the recipe and food made them feel. Then, using sticky notes, participants mapped out their instructions and ingredients and posted them on a wall, and the group would draw connections across recipes by finding common ingredients and meals that were tied to Mexican culture.

Initially, the group facilitators asked participants to write down their instructions on sticky notes using words, but quickly, the youths in the program incorporated visuals to illustrate their recipes and their journeys with the food. For example, the sequence of pictures in figure 4.8 represents how Alejandra, a middle school–age participant at La Escuelita, presented her recipe for making hardboiled eggs. Alejandra explained that her grandmother taught her to make hardboiled eggs to keep her healthy.

As evidenced in figure 4.8, Alejandra chose to use images rather than words to write down the technical instructions for her recipe. She used individual sticky notes to draw different-colored arrows that would guide users through the various steps in her recipe. When Alejandra presented her recipe to other members of the group, she translanguaged across Spanishes and Englishes, pointing to each sticky note so her audience could visually see the instructions she was verbalizing. In this way,

*Figure 4.8. Alejandra diagrams how to make hardboiled eggs.*

Alejandra provided access to her instructions through spoken words in Spanish and English as well as through her visual design. If members of the group didn't understand Alejandra's verbal explanation, they could follow her drawings and use the corresponding arrows to make sense of the different steps in her recipe.

In another recipe, Nubia, pictured in figure 4.9, illustrates how to make quesadillas, a dish she frequently makes with her mom. Nubia wrote numbered instructions in Spanish and then added short illustrations to demonstrate how to assemble her quesadillas.

As evidenced in these youths' creative recipe illustrations, when Escuelita participants are given the opportunity to move beyond alphabetic writing in their designs, they creatively layer Spanishes, Englishes, and visuals to make their work accessible to the broader group. Like the participants in the Diabetes Garage who emphasized the importance of visual communication in multilingual design, Escuelita participants used visuals to increase accessibility and support their verbal or written descriptions. This embracing of visual communication also extended as Escuelita participants worked to visually represent their community through the logo that would be depicted on all Escuelita materials, including their website.

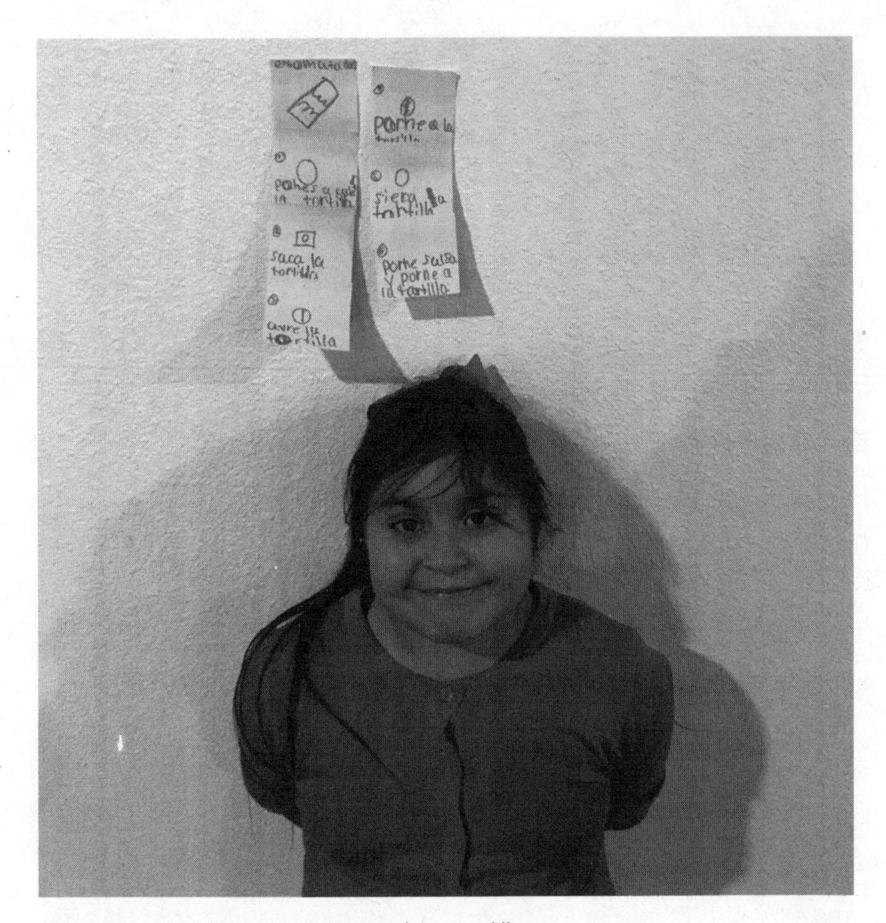


*Figure 4.9. Nubia documents how to make quesadillas.*

## TRANSLATING LOGO DESIGN

In addition to the design and structure of the website itself, one of the biggest conversations we had at La Escuelita concerned the design of our logo. Through several participatory design sessions, La Escuelita participants first developed prototypes of the logo that would go on the website. They began by brainstorming words that represent what they do, see, and feel at La Escuelita, which includes things like "*aprender*" (learn), "*investigar*" (research), "*convivir*" (share time together), and other fun things like "slime." Table 4.1 represents all the words participants associated with La Escuelita in our participatory design sessions. Based on these initial word associations, which included words and feelings in both Spanish and English, members of La Escuelita created initial sketches of the logo, one of which is depicted in figure 4.10.

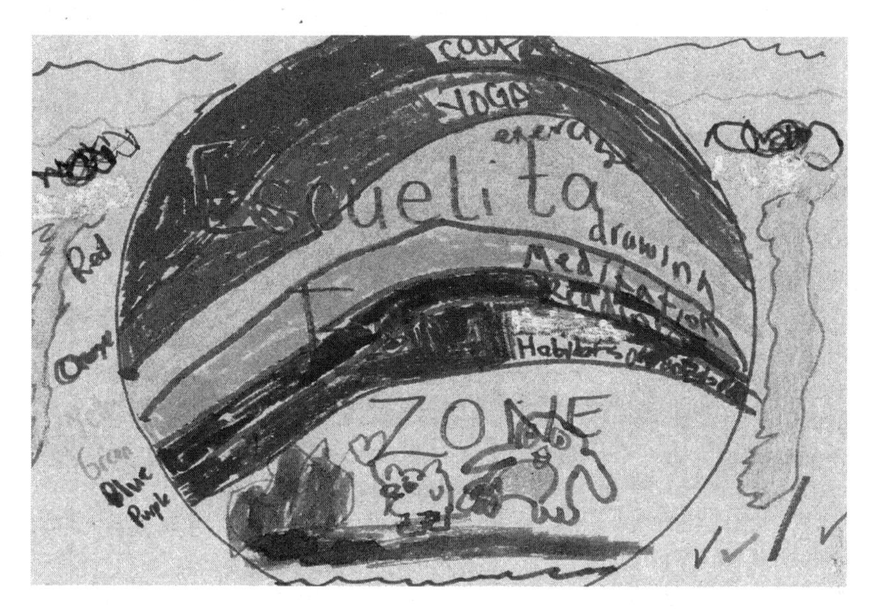


*Figure 4.10. Initial logo prototype.*

Table 4.1. Words participants associated with La Escuelita as we brainstormed a logo

| | |
|---|---|
| Aprender | Discover |
| Investigar | Slime |
| Convivir | Behave |
| Enseñar | Expresar |
| Creativo | Learn |
| Yoga | Respect |
| Succeed | Be safe |
| Books | Responsible |
| Fun | Kind |
| Inspirational | Art |

The logo prototype depicted in figure 4.10 was designed by Sergio, an elementary school–age participant at La Escuelita. Sergio titled his logo "Escuelita Fun Zone," and he included a circle with a rainbow at the top. When presenting this logo to the rest of the Escuelita team, Sergio explained that he included a rainbow because "La Escuelita makes me feel close to the stars, like the rainbow." He also listed activities we practiced at La Escuelita, such as "yoga," "meditation," "drawing," and "reading."

After making the initial paper prototypes, we engaged in follow-up design sessions with a visual designer who helped us create digital logos while also teaching our group about design. Although we began the prototyping session on paper, some youth participants at La Escuelita did have access to design software at school and chose to work on their prototypes digitally as well as on paper.

In collaboration with a visual designer, Patricia Flores-Hutson, members of La Escuelita then continued providing feedback on different

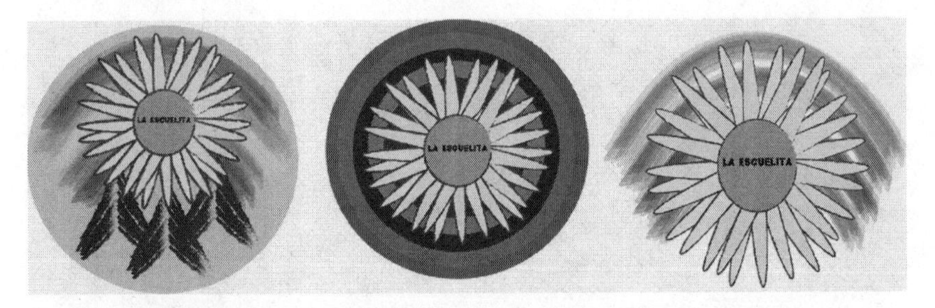


*Figure 4.11. Digital logo prototypes.*

iterations of the logo while also learning about design. Figure 4.11 depicts different versions of the logo that were created by the visual designer through several rounds of feedback with Escuelita participants. The logo on the far left of figure 4.11 is the one selected by Escuelita participants to represent our community. After Sergio made his comment regarding the role of the rainbow in his experience with La Escuelita, other members agreed that a rainbow should be incorporated into the logo design. As evidenced in figure 4.11, the visual designer incorporated elements of the rainbow into all three prototypes. However, when the designer presented these prototypes to Escuelita participants, the youths in the group insisted that the rainbow on the logo should follow the ROYGBIV format, the sequence of hues commonly described as making up the rainbow in the order red, orange, yellow, green, blue, indigo, and violet. As one of our participants explained, "The colors of the rainbow should be in the right order because it's science. At La Escuelita, we learn science and health." Through this discussion, it became evident that Escuelita participants were using their understanding of health, science, and language to make design recommendations while also considering the feelings and sense of community La Escuelita fostered. For example, the Escuelita youths chose the figure on the far left of figure 4.11 as their logo because the brown triangles underneath the flower represented the Franklin Mountain range in El Paso, a common landmark that made the Escuelita youths "feel like Escuelita is at home."

In transforming their ideas to visuals, participants at La Escuelita seemed less bound by static language categories and more focused on making information accessible to each other and to their audiences through the combination of Spanish, English, colors, and pictures. Likewise, participants in the Diabetes Garage emphasized the value of visual communication in language accessibility, noting that the Diabetes

Garage brochure would not be accessible to borderland residents without engaging, accurate visuals, regardless of how accurate the linguistic translations may be.

## TOWARD MULTILINGUAL DESIGN IN THE BORDERLANDS

At both La Escuelita and the Diabetes Garage, participants made contributions as savvy technology designers who understood the correlation between design and community values, knowing when information should be accessible to a larger audience and when information should be limited to the community itself. One of the biggest misconceptions technical communicators and technology designers might make when thinking about multilingual design is that *all* information in a particular platform should be translated and made accessible to *all* possible audiences. Yet as borderland participants demonstrate, sometimes ambiguity in meaning can make space for imaginative and innovative interpretations of design, and visual communication can aid in language accessibility when translating words into standardized languages may be ineffective or unnecessary. While participants at both sites certainly enact translation practices in their communication and design work, they are also keenly aware of the rhetorical power of words, images, visuals, and movements; and they negotiate these movements according to specific audiences, moments in time, and rhetorical situations. For these participants, words and language boundaries are parts of a broader creative technological toolkit, all of which function collectively in the communication of ideas and the creation of new technologies.

Scholars like Ofelia García and Li Wei (2015) illustrate the way bilingual and multilingual communities enact language not through rigid standardized systems but thorough fluid practices, which result in the constant blend and development of communication. In contexts such as the El Paso/Ciudad Juárez border, blending and mixing different varieties of Spanishes and Englishes is a common practice that has long historical roots. In businesses, classrooms, organizations, and other contexts, residents of the border deploy various linguistic practices to communicate, often extending beyond what may be conceived as the boundaries of standardized English or Spanish. While this fluidity of language might be commonly accepted in everyday contexts, the long history of violence and stigma against "inaccurate" or "improper" language (either English or Spanish) leads borderland residents to have particular attitudes and hesitancies toward their own blended and fluid languaging practices (Hidalgo 1986). For example, in conversations that took place during the participatory translation focus groups in the Diabetes Garage, participants distinguished their own language practices from those of "the people that are actually from . . . the Mexican side" or the people who "were raised with the proper Spanish." As Peñalosa (1980) explains, the "unity" of the borderland population is "contested by some who prefer to distinguish the 'assimilated' from the 'nonassimilated,' or Mexican-born 'Mexicans' from US-born 'Chicanos,' or Spanish Americans from Mexican Americans" (2). Yet, as Peñalosa (1980) clarifies, "Chicanos are an extremely heterogeneous population that varies along every conceivable social dimension" (2).

As a technical communicator invested in helping to design accessible materials for and with this borderland community, recognizing the various elements embedded in this particular multilingual technical communication experience allowed me to understand that in soliciting and applying community stakeholders' feedback into the design of health-related materials (e.g., a brochure about diabetes treatment and prevention, a website for a health literacy after-school program), I should also take into account the socio-cultural and rhetorical contexts in which this feedback is provided. For this borderland community, long-established language ideologies should be considered in the design of bilingual materials, where the job of our research and design team extended from merely providing localized content in both Spanish and English to making decisions about whose languages we would or should value in our final designs.

For example, in designing the Diabetes Garage brochure, based on participants' suggestions, our research team made translation recommendations that incorporated borderland language into the design. We recommended, for example, the use of the term *checa* instead of the more standardized *revisa* in the design of the finalized brochure, since checa seemed to resonate with most of the participants. However, once the brochure was sent off for revision and printing at the university communications center that would be producing and printing the brochures, the translation of "checa" was reversed to "revisa" by the university communications team. Thus, as evidenced by this brief example, even in instances when researchers and designers seek to design and localize translations, public entities (in this case, the local university) still operate under standardized, white/Western language ideologies—those who position non-standardized discourses as inferior or "offensive" in comparison to their standardized counterparts. For similar reasons, our Escuelita research team chose to build the Escuelita website as a standalone site on a content management system (CMS) outside of our university site, even though the university could have provided web space free of charge. In this particular case, our team chose to pay for the CMS to keep control of the way we could represent our language and design practices. In both contexts, intersectional, interdependent methodologies can help researchers understand the various connections among language, power, and access that are always at play in design scenarios, particularly when working with communities of color in this borderland context.

## CONCLUSION

From a translation and localization perspective, the data narratives I presented from La Escuelita and the Diabetes Garage can render several implications about how communication designers can design for and with borderland communities. For example, findings point to the fact that technologies for borderland communities should mirror the languaging practices of borderland residents, resulting in websites that are, as La Escuelita participants suggested, not divided into two sections, one in Spanish and one in English. In addition, participants in both La Escuelita and the Diabetes Garage emphasized the value of visual communication in multilingual design, which also echoes long-standing research about the importance of visuals in health communication (Bloom-Pojar and DeVasto 2019). However, established infrastructures for producing these designs, including university systems and formalized learning and teaching spaces, do not always welcome these fluid approaches to language representation. As evidenced by the discussions held with the Diabetes Garage participatory translation focus groups, borderland residents are critically aware of how these formal infrastructures for information design police language representation. For this reason, borderland residents themselves may resist embracing fluid approaches to language and translation, since they understand, through embodied lived experiences and through their constant navigation of racial and linguistic oppression, that while borderland residents may have extensive expertise in and prefer borderland Spanishes and Englishes, formal design spaces are not always welcoming of this expertise and of the communities that embody fluid, non-standardized languaging practices.

As technical communicators, we have a responsibility to embrace fluid notions of language used by communities such as the borderland residents depicted in this project. At the same time, as researchers who publish about technical communication and information design, we also have a responsibility to shift broader conversations and infrastructures that prevent fluid language practices from being embraced and welcomed in formal learning, teaching, and design spaces. While communities like the borderland residents depicted in this chapter have extensive linguistic, cultural, and technological expertise, this expertise continues to be erased from the infrastructures responsible for designing and disseminating information to the very communities that could benefit from more localized approaches to information design.

At the core, participants in La Escuelita and the Diabetes Garage illustrate the value of intersectional, interdependent approaches to language and translation in design. While many researchers across fields may be in the practice of conducting focus groups and other participatory design activities like those presented in this chapter, an attunement to the historical underpinnings of language difference can help researchers understand the motivations for some design recommendations while also helping researchers co-design spaces where established rules and protocols for language practice can be expanded to rely more directly on localized community preferences. Co-designing materials in community with borderland residents requires more than simply asking for design preferences; it requires establishing multilingual experiences that foster open conversations about language stigma and raciolinguistic ideologies that continue to shape how information is broadly disseminated, particularly in digital spaces. As designers, healthcare professionals, and community activists continue to design and share tools and platforms in this powerful borderland context, more attention should be paid to the ways borderland communities—and for the purposes of this study, communities that live, study, and work in and between El Paso, Texas, and Ciudad Juárez, Chihuahua—read, understand, and conceptualize information across platforms and languages. Intersectionality and interdependency as methodologies should be applied not only in individual local projects but also in broader collaborations among various stakeholders responsible for creating and sharing information with and for marginalized publics.

In this chapter, I provided some examples of how methods such as participatory translation and collaborative design can illuminate the linguistic, racial, and cultural ideologies that underpin contemporary communicative contexts and spaces. While the Diabetes Garage and La Escuelita represent various different demographics, age groups, and genders, the emphasis on "formal" versus "informal" language and the leveraging of visual design resonated across participants in these collaborative design projects. In the context of these borderland projects, my own linguistic expertise in Spanish and my positionality as a Latina provided some opportunities to connect with participants and discuss concepts in depth without much linguistic limitation, since all parties included in these projects felt comfortable communicating in Spanish, English, or a combination of both languages. In chapter 5, I introduce a participatory design research site with far different linguistic dynamics in the South Asian context of Kathmandu, Nepal. As I continue to move across research sites throughout the remainder of this book, I continue to highlight the importance of core concepts like collaboration, translation, community trust and expertise, and an attunement to linguistic positionality that can inform technical communication research in international and transnational contexts. Before moving on to the next chapter, I conclude this chapter with a reflection from Judith Hernandez-Rivera, an Escuelita participant who is now in her first year of college. In the reflection, Judith describes the community space created by our Escuelita family.

## A REFLECTION ON LA ESCUELITA, BY JUDITH HERNANDEZ-RIVERA

Since I started going to La Escuelita, everything feels different in many ways. Before attending this after-school activity, I was struggling a lot with my identity and how I viewed myself. I struggled a lot and had a lot of problems with school and problems with a lot of people around me. In school I was being bull[ied] and picked on; many kids would call me fat, ugly, even stupid during class. My teachers didn't really do much about it. In La Escuelita I met so many people that helped me see who I really am like Laura, William, Victor, and many of my "classmates," especially Heidi. Heidi and I would guide and help the younger classmates. It was so fun and relaxing. They helped me overcome many of my doubts about my future career and about myself.

At first when my mom forced me to go to La E[s]cuelita, it was just a waste of time in my point of view. I would always be mad and didn't participate at all. I would see it as a punishment because I was being forced to go and tolerate people that I didn't care for at the time. I would just sit there and stare into space. But I remember Lucía, one of the teachers that would be there with us, offering me a drink and started asking me questions about myself and school. At first, I felt weird about it because nobody was interested in me, and I guess that little attention made me want to reach out and participate more and actually get to know them more and see what was La Escuelita about. I loved the activities [and] always having my attention and having that help that I was scared to ask for at the time. I started going to La Escuelita maybe when I was like 12 or 13. Currently, I'm 18 and through the years La Escuelita and everyone that has been there has seen me grow and develop as the person that I'm now. I went from a sad and cold teenager to an outgoing and dedicated person, or that's what most of my teachers tell me I am. I really encourage people to participate more in their communities. If the[y] can't change the way people think, you have that amazing experience and find many people that see how unique [you are] and that are willing to help you.

# 5
# USER EXPERIENCE AND PARTICIPATORY DESIGN IN KATHMANDU

From 2016 to 2019, while I was living in El Paso and working on the projects described in chapter 4, I was also teaching at a university in this borderland region, where I had the opportunity to meet international graduate students from across the world. The undergraduate population at this institution is approximately 80 percent Latinx, with an additional 7 percent of the student population identifying as Latinx students who are Mexican nationals and who live in Ciudad Juárez, Chihuahua, Mexico, and commute to the campus in El Paso to attend school.

The demographics of international students at this university are not reported at the graduate level. However, as a graduate faculty member teaching in a PhD program in rhetoric and writing studies at this institution, I noticed upon my arrival that unlike the predominantly Latinx undergraduate population, a majority of my PhD students identified as international students with home countries that included Nepal, Mexico, Ukraine, Ghana, and Cameroon, among others. In each of the six graduate seminars I taught at this institution, for example, the vast majority of students identified as international students from the aforementioned countries; in courses that ranged from eight to fifteen students, I never had more than three domestic students in a single course.

In addition to the demographics in my courses, as a member of the PhD program application review committee, I also noticed that each year, we received multiple applications from Nepali students who wanted to study rhetoric in the US, and each year, several members of our incoming PhD cohort were Nepali international students. This representation of Nepali students in the graduate cohort echoed national trends. According to the 2018 Open Doors Report on International Educational Exchange (IEE) data released by the IIE and the US Department of State Bureau of Educational and Cultural Affairs, the number of Nepali students enrolling in graduate school in the US

https://doi.org/10.7330/9781646422760.c005

had steadily increased over the previous twelve years, reaching a peak during the 2017–18 academic year (US Embassy in Nepal 2018). The reasons for this increase in Nepali student enrollment in US institutions are varied and complicated, but the initial takeaway for me as a faculty member in a PhD program was that I wanted to find ways to localize our program content and structure to echo our student population. This was particularly important to me as a faculty member who taught the digital rhetoric, technology, and technical communication courses in the program. Orienting to technology and technical communication through the social justice–driven frameworks presented in this book, I wanted to find ways to expand our program's definition of technology and effective communication design, and I knew that to do so, I needed to learn from our Nepali students. In addition, as I began my career on the tenure track in this context, at this institution with brilliant students from so many places, I kept going back to a question several people had asked regarding my previous research (including the work presented in chapter 4): "Your work on multilingualism and technology with Spanish/English speakers is really interesting, but have you ever worked in multilingual contexts that include languages you *don't* speak?"

I resisted this question for several years because I believed (and still believe) that speaking the same language(s) as your participants can open up important possibilities during the research process. I'm also critically aware of the damage researchers can cause when working with multilingual communities of color without having the cultural expertise and experience to engage in research through justice-driven orientations that intentionally avoid fetishizing cultural and linguistic practices. My answer to the question "Should researchers do work with multilingual communities of color who speak language(s) the researcher(s) does not understand?" is still complicated and context-dependent. However, in this chapter, I document how I moved from doing multilingual technical communication work with Spanish-/English-speaking communities to working with communities with which I do not share linguistic expertise. In particular, I illustrate the many tensions, decision-making points, affordances, and constraints that emerged from my collaboration with Nepali students who are affiliated with the South Asian Foundation for Academic Research (SAFAR), an academic research center in Kathmandu, Nepal. Through this discussion, I hope to provide some answers and possibilities for other technical communication researchers seeking to do work with multilingual communities of color through justice-driven orientations.

## ABOUT SAFAR

In 2019, I had the opportunity to visit Kathmandu to co-facilitate (along with then PhD student Bibhushana Poudyal) a three-week workshop series on participatory design and digital humanities at SAFAR (see Poudyal and Gonzales 2019). Together, we decided that the workshop series would focus on participatory design and user experience while also highlighting previous work on digital humanities conducted by SAFAR. Under the leadership of Professor Arun Gupto, director of SAFAR, we were awarded funding for this project through the Fulbright Specialist Program, and we also received support and funding from the Coalition of Women of Color in Computing to trace the process of developing participatory design practices with this group of student participants in Kathmandu.

SAFAR is an independent research center in Kathmandu described as an "interdisciplinary and comparative discursive platform" (http://www.safarsouthasia.org). According to the center's website, SAFAR's mission is "to create a venue for dialogue and debate between international scholars working on South Asia." The projects SAFAR develops are intended to bring multiple perspectives to South Asian studies through various methodologies, many of which incorporate the design and distribution of digital platforms and technologies. Countering traditional narratives about a "lack" of digital literacies in South Asian contexts, SAFAR is a space of collaborative innovation that leverages digital technologies to engage in critical inquiry and community-driven research. Thus, SAFAR graciously welcomed Bibhushana and me to lead workshops related to digital technologies that would benefit students' individual research goals as well as the overall mission of the research center.

## ABOUT THE WORKSHOPS

Over the course of three weeks, we led digital writing workshops at SAFAR through a participatory design perspective. The goal was to help our student participants, who were at various levels of their graduate careers, become more familiar with digital research tools, methods, and conversations. In addition, the workshops were intended to help students develop their own digital projects related to their specific research interests. A participatory design approach was used in developing these workshops to (1) foster engagement from students, (2) encourage students to incorporate their own backgrounds and interests into workshop activities and discussion, and (3) allow students with various levels of experience working in digital environments to work and learn together. Throughout and beyond the workshops, students at SAFAR undertook their own projects that highlighted specific research interests and goals.

Before moving on to discuss my own positionality and orientation to this project, I include an introduction to and summary of this chapter written in Nepali for Nepali readers by a workshop participant, Pragya Dahal. The full text of Pragya's summary is posted here in Nepali, and an English translation is provided as a footnote.[1] Pragya provided her

1. The number of Nepali students in [the] USA used to be quite low, but it has distinctively increased in these few years. The author, as a PhD faculty member, has always wanted to find ways to localize her program's content and structure so that her students['] experiences also get reflected in the courses designed. The reliance on English as a primary form of communication is common and documented through a long history of linguistic imperialism in Nepal and all of South Asia. The author shares her experience that in the international seminars she has conducted she didn't get [a] chance to have all students who represent the original locality where they were conducted. She also believes that speaking the same language(s) as your participants can open up important possibilities during the research process.

A workshop series was conducted by the author's team that basically focused on participatory design and user experience, taking into account the local language, Nepali, here in Kathmandu. The three-week workshop [was] intended to bring multiple perspectives to South Asian studies through distinct methodologies that assisted students like myself in becoming more familiar with digital research tools, methods[,] and conversations. Following a participatory design approach, the workshop [was] intended to foster our engagement, encourage us to incorporate our backgrounds and interests into workshop activities and overall discussions, and create a co-learning and sharing environment with students from various backgrounds.

In order to break the barriers that might be created by the English language, translation was used as one of the major tools of the workshop.

English language in Nepal emerged [quickly] after the neighboring British colonization in India[,] which sometimes is recognized as the easiest way of influence and international recognition and opportunity as scholars and academics. During the workshop, the "space" created by the Nepali students' inability to directly type Devanagari script in the Western-designed keyboards was noticed, and it provided an opportunity for discussion. As our workshop focused on creating such digital spaces in education, mass media[,] and other domains of language use, this lack of Devanagari script remained one of our major topics of discussion. This indicated the largely dominant nature of [the] English language over the other languages and scripts, which has led to Nepali speakers typing in Nepali using Roman script.

This indicates the lack of usability research conducted with Devanagari keyboards, which makes these keyboards inaccessible for users with limited mobility. Despite this huge gap, many Nepali speakers have developed innovative tools, both digital and material, to maintain their language and writing systems. Whereas sometimes Nepali itself is positioned as a dominant language used to erase Indigenous languages, the Indigenous language users have also developed their own typing technologies to aid in language preservation and revitalization. This all represents the value of language positionality during collaborative research.

During the workshop, the participatory design process was applied to make space for participants' views and feedback. In order to keep us engaged, participatory

own translation for her summary. Following Pragya's summary, I discuss my own positionality and the structure of this project.

## PRAGYA'S REFLECTION

मेरो समिक्षा :

केही वर्ष यता अमेरिकामा नेपाली विद्यार्थीहरुको संख्या विशिष्ठ रुपमा बढेको छ । एक विद्यावार ीधि संकाय सदस्यको नाताले लेखकले आफ्ना विद्यार्थीहरुका लागि बनाईएको पाठ्यक्रममा उनीहरुलाई नै प्रतिबिम्बित गर्न सँधै पाठ्यक्रम सामाग्री र संरचनाको स्थानीयकरण गर्न तरिकाहरु खोज्नुभएको छ । नेपाल लगायत सम्पर्ण दक्षिण एसियामै अंग्रेजी भाषाको लामो भाषिक साम्राज्यवादले जरो गाडे को र यसको दस्तावेजीकरण गरिएको सन्र्दभमा अंग्रेजी भाषा प्रतिको निर्भरता सामान्य हो । आफूले आयोजना गरेका अर्न्तराष्ट्रिय सेमिनारहरुमा सोही स्थानीय समुदायको प्रतिनिधित्व गर्ने विद्यार्थीहरुको संख्या निकै न्यून रहेको तीतो अनुभव लेखक व्यक्त गर्दछिन् । सहभागीहरुले प्रयोग गर्ने भाषा (हरु) प्रयोग गर्दा मात्र पनि अनुसन्धान प्रकृयाका विभिन्न सम्भावनाहरु खुल्दै जाने विश्वास लेखकको छ ।

लेखकको समूहले आयोजना गरेको कार्यशाला श्रृंखलामा मूलत : काठ्माडौंको स्थानीय भाषा मानिने नेपाली भाषाको ख्याल गर्दै सहभागी र प्रयोगकर्ताका अनुमक्मा केन्द्रित रही कार्यशालाको

---

translation was done, applying the usability tests where we visited various websites and Google translations were performed. After this[,] discussions were had about the many ways that digitalization, including the use of digital translation software as well as word processing software, changes the representation of language and culture in both material and metaphorical ways.

This activity led to a mass realization that some type of reshaping and distortion might happen when our identities are conveyed through digital platforms like professional websites in specific content management system[s] like Wordpress.

This was followed by the participants handling the sticky notes that had various topics and words which we needed to define and provide translations individually[,] later making connections between the sticky notes and finding patterns in them.

The workshop was aimed at highlighting students' innovative practices, language skills[,] and backgrounds as opportunities for technological innovation. Also, a yarn game was played in order to know the status of our digital identity and know the research interests of the participants to know more about how we all were connected and how that connection could be continued in/through the digital platforms. This yarn activity was new to all of us[,] which encouraged us to share ideas and create a shared space to build a collective community.

Additionally, all of us designed our own websites that included our interests and could also contribute to the community. I, as a participant of the workshop, redesigned the website that I maintain for the organization where I work as an advocacy and communication officer. Later, I also learned a new tool for web analytics and ways to apply it in our website to further track visitors.

This workshop particularly changed our lenses regarding our positionality of language[,] which is fluid in nature (that can change on the basis of time and space). It also helped illustrate how translation activities and language positionalities help us illuminate Western English dominant frameworks. Also, the technologies through which communication happens in digital spaces matter a lot. The major motifs of the various sessions conducted in the workshop included the various shared objectives. This workshop also enabled us to create free platforms for publishing our own websites and working collaboratively for scholarly article publication.

मुख्य उद्देश्य नगुम्ने गरी गरियो । उक्त ३ हप्ते कार्यशालाको मुख्य उद्देश्य नै दक्षिण एसियाली अध्ययनमा म जस्ता विद्यार्थीहरुलाई विभिन्न दृष्टिकोण र भिन्न प्रकृयाहरुको प्रयोगबाट "डिजिटल" अनुसन्धान उपकरणहरु, विधिहरु र कुराकानीहरुका बारे परिचित गराउनु थियो । विभिन्न परिवेशलाई प्रतिनीधित्व गर्ने विद्यार्थीहरुलाई सहभागितामूलक "डिजाइन"को दृष्टिकोण अनुसरण गराउँदै कार्यशालाले हाम्रो संलग्नतालाई बढावा दिन , हाम्रो पृष्ठभूमि र रुचिहरुलाई कार्यशालाका गतिविधिहरु र समग्र छलफलहरुमा समाहित गर्न एक सह-शिक्षा र साझा वातावरण सृजनाको प्रणाली अपनाएको थियो ।

सहभागीहरुको अंग्रेजी भाषा बुझ्ने असमर्थताले उत्पन्न गराउन सक्ने बाधाहरुको मध्यनजर गर्दै अनुवादलाई कार्यशालाको महत्वपूर्ण उपकरणका रुपमा प्रयोग गरिएको थियो । विद्वान एवं शिक्षाविदहरुका अनुसार, भारतको बेलायती उपनिवेशले सहज रुपले अंग्रेजी भाषाको प्रभाव , अन्तर्राष्ट्रिय मान्यता र अवसरहरुको सृजना गर्यो । कार्यशालाका क्रममा नेपाली विद्यार्थीहरुले सहज तरिकाबाट अंग्रेजी किबोर्डमा देवनागरी लिपीको प्रयोग गर्न नसकेका कारण उत्पन्न भएको "स्पेश" को बारे पनि सबैको ध्यान आकृष्ट भएको थियो । किनकी हाम्रो कार्यशाला शिक्षा , संचार माध्यम र भाषा प्रयोगका अन्य क्षेत्रहरुमा "डिजिटल स्पेश" को सृजना गर्ने कुरामा केन्द्रित थियो , देवनागरी लिपीको कमी हाम्रो छलफलको मुख्य विषय बन्यो । नेपाली भाषीहरुले नेपाली भाषा समेत रोमनमा "टाइप" गर्दछन् , यसले अंग्रेजी भाषा , अन्य भाषा र लिपीहरु प्रति अत्यन्त हावी भएको समेत इंकित गर्यो । साथै यसले देवनागरी भाषाको "किबोर्ड" को प्रयोगयोग्य अनुसन्धान खड्किएको संकेत गर्दछ भने सिमीत गतिशीलता भएका प्रयोगकर्ताहरुका लागि यो "किबोर्ड" पहुँचयोग्य नभएको तथ्य पनि पुष्टि गर्छ । यति ठूलो अन्तरका बाबजूत पनि धेरै नेपाली वक्ताहरुले आफ्नो भाषा र लेखन प्रणाली कायम राख्न नयाँ "डिजिटल" र भौतिक उपकरणहरुको विकास गरेका छन् । कहिलेकाहीँ- अंग्रेजी भाषा नेपाली र अन्य भाषाहरु माथि हावी भए जस्तै ) नेपाली भाषा स्वंय पनि आदिवासी जनजातीका भाषाहरु माथि हावी भएको पाईन्छ फलस्वरुप आदिवासी जनजाती भाषाहरुका प्रयोगकर्ताहरुले पनि आफ्नो भाषाको संरक्षण र पुर्नउत्थानका निमित्त आफ्नै "टाईप" गर्ने प्रविधीको समेत विकास गरेका छन् । यो सबैले सहयोगी अनुसन्धानका क्रममा भाषा स्थिती मापन गर्ने तरिकाहरुको प्रतिनीधित्व गर्दछ ।

कार्यशालाका क्रममा तालिमकर्ताले आफ्नो ज्ञान आदान प्रदान गर्दा विद्यार्थीहरुको विचार र पृष्ठपोषणमा असर नपारोस् भन्ने उद्देश्यले सहभागितामूलक "डिजाइन"को अवधारणा अबलम्बन गरिएको थियो । विद्यार्थीहरुलाई व्यस्त राख्नका निमित्त सहभागितामूलक अनुवाद प्रकृयालाई प्रयोगकर्ता परिक्षण विधी अप्नाई विभिन्न "वेभसाईटहरु" हेर्ने र "गुगल" अनुवाद समेत गरिएको थियो । यस पश्चात "डिजीटलाईजेशन" का विभिन्न तरिकाहरु जस्तै "डिजीटल अनुवाद सफ्टवेयर" , "वोर्ड प्रोसेसिङ्ग सफ्टवेयरले दूबै भौतिक र "मेटाफोरिकल" तरिकाबाट भाषा र संस्कृतिको प्रतिनीधित्व परिवर्तन गर्छ भन्ने विषयमा छलफल गरियो । यस गतिविधिले एक सामूहिक अनुभूति दियो कि जब हाम्रो परिचय पनि "डिजीटल" माध्यमबाट प्रस्तुत गरिन्छ यस्तै प्रकारको पुन : आकार र विकृतीको सामना गर्नुपर्दछ जस्तै "वोर्डप्रेस" जस्ता विशेष सामाग्री व्यवस्थापन संग सम्बन्धीत वेभसाईटहरु मार्फत ।

यस पछि सहभागीहरुका हातमा विभिन्न शीर्षकहरु जस्को छुट्टा छुट्टै अनुवाद र व्याख्या गरिन'पने "स्टिकी नोट" थमाईयो , त्यसपछि ती फरक फरक "स्टिकी नोट" को बीचको सम्बन्ध पत्ता लगाईएको थियो ।

उक्त कार्यशाला मुख्य रुपमा विद्यार्थीहरुको नवीन अभ्यासहरु ,भाषा सीपहरु , र प्राविधीक नविनताको विकासका अवसरहरुको खोजी गर्न लक्षित थियो । साथै एक उनको डल्लो को खेल जसले हाम्रो "डिजीटल" परिचयको स्थिती , हामी एकअर्का संग कसरी सम्बन्धित छौं , र यो सम्बन्धहरुलाई

"डिजीटल" माध्यमका मार्फत कसरी निरन्तरता दिने भन्ने बारे खेलाईएको थियो । त्यो उनको डल्लाको खेल हामी सबैको लागि नयाँ थियो जसले हामीलाई आफ्ना विचारहरुको आदानप्रदान गर्दै एक साझा "स्पेश"को सिर्जना गरी सामूहिक समुदायको निर्माण गर्न उत्प्रेरित गर्यो ।

साथै हामी सबैले आफ्नो लगाब भएको विषय जसले समुदायलाई पनि योगदान पुऱ्याउन सक्छ , त्यस्तो विषय छानी "वेभसाईट" सृजना गयौं । म उक्त तालिमको एक सहभागीका हैसियतले आफूले सूचना तथा पैरवी अधिकृतको रुपमा कार्यरत रहेको संस्थाको संस्थागत "वेभसाईट" "पुन : डिजाइन" गरें । पछि मैले एउटा नयाँ प्रविधी समेत उक्त तालिम बाटै सिकें जसलाई "वेभ एनालाईटिक्स्" भनिन्छ , मैले यो प्रविधीबाट हाम्रो "वेभसाईट" हेर्ने दर्शकहरुको अध्ययन गर्ने मौका पाएं ।

यस कार्यशालाले विशेषत हाम्रो भाषाको स्थिती बारे हाम्रो दृष्टिकोण परिवर्तन गर्यो । हाम्रो भाषाको स्थिती परिवर्तनशील छ , जुन समय र ठाउँको आधारमा परिवर्तन हुन सक्दछ । यसले अनुवाद कृयाकलाप र हाम्रो भाषा स्थितीका माध्यमबाट प्रमुख पश्चिमा अंग्रेजी संरचनाहरुको उजागर गर्न सकिन्छ भन्ने पनि चीत्रण गर्यो । साथै प्रविधीहरु जसको माध्यमबाट "डिजीटल स्पेश" जहाँ सूचना आदान प्रदान हुन्छ , निकै महत्वपूर्ण हुन्छ । कार्यशालामा संचालन गरिएका विभिन्न सत्रहरुको छुट्टा छुट्टै महत्व र भूमिका रहेको थियो । "ल्यान्डस्केप विश्लेषण" ले मुख्यत : "वेभसाईट" सृजनाको अवधारणा विकासमा ठूलो योगदान पुऱ्यायो । "क्रिटिकल डिजीटल ह्यूम्यानिटिज" का बारेमा गरि एको विचार आदान प्रदानले मानिसहरुको ज्ञान , उपायहरु , स्थानहरुको "डिजीटल प्रतिनीधित्व" विश्लेषणात्मक तरिकाले हेर्न सिकाउँछ । "डिजीटल पहिचान सृजना" गर्नुले "डिजीटल प्ल्याटफर्ममा" हाम्रो व्यक्तित्वका बारे स्पष्ट पार्दछ । यसले हामीलाई सामाजिक संजालमा संलग्नता, प्रतिनीधित्व, र व्यक्तिगत वा पेशागत "वेभसाईट डिजाईनमा" समेत सहयोग गर्दछ । यस कार्यशालाले हामीलाई हाम्रा आफ्नै "वेभसाईटहरु" बनाउने निःशुल्क स्थानहरुको सृजना गर्न र सहकार्य गर्दै बौद्धिक लेख र चनाहरु प्रकाशन गर्न समेत सक्षम बनाएको छ ।

धन्यवाद ।

## RESEARCHER ROLE AND POSITIONALITY

While many researchers across fields, including technical communication, are now in the common practice of acknowledging their positionalities in relation to their research participants and contexts, I find it important not only to recognize my own positionality in a specific context or project but also to note how this positionality shifts and changes as we move across communities and settings. In the research projects documented in chapter 4, for example, as I worked with Spanish-speaking communities in El Paso, my identity as a Latina and my linguistic history as a Spanish speaker afforded me specific connections with my research participants. In the context of Nepal, however, my positionality shifted in various directions, influenced by my lack of ability to speak the same languages as my participants and my lack of cultural knowledge and expertise related to Nepal. What remains across contexts is my white privilege, as well as my positionality as an academic from a Western institution who can speak and write standardized American English.

In the context of this project at SAFAR, I relied on my expertise as a language and translation scholar as well as my experience as a technical communicator and technology designer to work toward fostering conversations that highlighted rather than ignored our group's linguistic, racial, and cultural differences. Approximately twenty-five students from Nepal and one student from India participated in this project. The twenty-five participants from Nepal as well as the workshop co-facilitator spoke various languages, including Nepali, Hindi, English, and Rai, an Indigenous language in Nepal. Due to my linguistic limitations, the group activities took place largely in English. This reliance on English as a primary form of communication is common and is documented through a long history of linguistic imperialism in Nepal and all of South Asia (Shrestha 2016; Pandey 2020), which I will expand on in the next section. The centralizing of English in a Nepali context, not only in the workshops I co-facilitated but also in Nepal's long connection to British colonization, was a critical yet overlooked element that impacted the way our workshop community engaged with participatory design.

## ENGLISH IN NEPAL

While researchers in technical communication and other fields may increasingly acknowledge our positionalities as researchers, particularly when working with communities of color and international communities, how often do we take the time to recognize the positionalit(ies) of the language(s) we use to do our work? When doing research with immigrant communities in the US, for example, how often do we as technical communicators rely on our participants' knowledge of English to hold conversations, despite the fact that English may be their second or third language? Our research protocols as US academics and industry researchers, including our methods and methodologies, Institutional Review Board (IRB) forms, and consent practices, as well as our general research training are so deeply rooted in standardized white American English that many of us don't take the time to question why and how the presence of this standardized form of communication inherently and automatically impacts the types of conversations we can have with participants and the environment in which our work takes place.

In my work in Kathmandu, the reliance on English as a primary form of communication placed the labor of translation on the most marginalized members of our workshop space while providing me as the Western researcher with the most comfort and flexibility. Defaulting to English as the "lingua franca" of our classroom space, whether intentionally or not, perpetuated a long, ongoing, and complex history of linguistic colonization in Nepal.

In "English in Nepal: A Sociolinguistic Profile," Shyam B. Pandey (2020) traces the history of the English language and its implementation in Nepal through British colonization. As Pandey (2020) explains, more than 100 languages are spoken in Nepal, and "English is the second most frequently used language across various domains throughout the nation" (1). The use of English in Nepal is rooted in British colonization of South Asia as well as in Nepal's caste system. In 1801, following British intervention in internal conflicts among Nepal's rulers, Nepal signed a treaty with the East India Company that "allowed the British to have permanent residence in Kathmandu," thus perpetuating colonial influence and the influence of English on institutionalized processes and procedures (Pandey 2020, 2). Later, in the 1850s, Prime Minister Jang Bahadur Rana established the "first English medium school" in Nepal, known as "the Durbar School, which was set up within the Thapathali palace" and "was exclusively for the large Rana family" (Pandey 2020, 3). As Krishna Bista, Shyam Sharma, and Rosalind Latiner Larby (2019) explain, English in Nepal later "evolved from being a secret means for the ruling oligarchs to gain connection with the world outside, to an aspiration for greater social capital among the upper class and caste, to, finally, becoming a rather unrealistic aspiration for the middle class" (138). In the context in which I was working during my time in Nepal, English was already centralized as a means to higher education and was sometimes perceived as a way for students to gain international recognition and opportunity as scholars and academics. Furthermore, although English is a dominant language used in academic spaces in Nepal, Nepali itself is also sometimes used as a "lingua franca" in Nepali academic spaces. While students, including the participants in the workshops I describe in this chapter, may speak multiple Indigenous languages of Nepal at home, Nepali and English are sometimes seen as superior in academic spaces. Thus, linguistic colonization in Nepal is not limited to English alone.

My presence as a Western academic teaching in Nepal was also normalized through the initial agreement between Nepal and the East India Company, where Western-based academics became positioned as intellectual experts with a claim to reside in Nepal. As Bista, Sharma, and Larby (2019) continue, "In this brief history of Nepal's higher education, politics and political economy remain the most potent shaping forces," also contributing to the increased number of Nepali students seeking graduate education in the United States (135). Thus, as I interrogated my own positionality in this project in conjunction with the positionality of language in this research space, it became more evident that all of the pieces making this project possible were interconnected and historically rooted. When technical communicators go into communities to conduct research and collaborate, the access we have to communities and the practices through which we conduct our work are often influenced by long-standing infrastructures of privilege and oppression. This also applies to the technologies we use in participatory design research.

### NEPALI UNICODE

Recognizing linguistic positionality continued to gain importance in this project as the students and I began working together during our workshops. Our workshops began each day at 1:00 p.m., with students working on some element of their digital projects on their own computers. Early in one of our workshop sessions, I walked around the room and noted that some of the participants typing content in Nepali were using open-access tools such as the Nepali Unicode converter (https://www.nepali-unicode.com) to facilitate their typing. Unicode is a standard set of symbols used to consistently encode, represent, and preserve various scripts of the world in digital formats. A Nepali Unicode converter transforms Roman script into Devanagari script, the script used to type in Nepali and Hindi. Completely oblivious to Nepali Unicode converters and their uses, I asked students in the workshop to tell me more about this tool and why they were using it. Through our conversations and additional research, I then learned that the traditional script, Devanagari, is an alphasyllabary script that is phonetic in nature and that includes twelve vowels and thirty-three consonants. While some Nepali speakers still learn to write and read Devanagari script, over the last thirty years—particularly with the incorporation of typing into schooling and other forms of communication—Devanagari script is sometimes replaced when users write in Nepali using a Roman alphabet. For example, a common Nepali greeting for hello is नमस्ते, which is written as "namaste" in Roman script. While the traditional written representation of नमस्ते is in Devanagari script, in contemporary contexts like school assignments that use word processing software as well as social media and other texting platforms, many Nepali speakers, including some of the SAFAR workshop participants, are in the common practice of using Roman rather than Devanagari script to type. As one of my students' daughters commented when I asked her about Devanagari script, "Devanagari is something we use when we are making memes or saying something funny or creative. For everything else we type on the computer, we use Roman script."

While the preference for Roman script may be dismissed as a simple matter of choice and while many workshop participants claimed that typing in Roman script is "easier" or "faster," the reasons for this speed and efficiency are also rooted in discriminatory practices linked to the development of keyboards. In their chapter "Unicode and Nepali," Chetan Prajapati, Jwalanta Deep Shrestha, and Shishir Jha (2008) link the "absence of standardization of Nepali Keyboard layout" to the diminishing use of Devanagari script in digital contexts. As the authors explain, "Unicode provides a unique number for every character used in the computer, no matter what the platform, no matter what the program, no matter what the language. This completely minimizes the conflicts and data corruption caused by the incompatible coding system" (2). While many languages and their scripts are incorporated into current Unicode charts, Nepali Devanagari script was not assigned to Unicode standards in a timely fashion; because of this, Nepali speakers began typing in Nepali using Roman script. As Prajapati, Shrestha, and Jha (2008) explain, "Traditionally Nepali fonts such as Himali, Preeti, Kantipur, Sama[,] etc. has [*sic*] been developed and used to fulfill the need of documents required in Nepali language. All these typefaces, though they use the Devanagari font as their base, have different coding system[s]. This brings about a lot of complication in downloading the documents from one pc to [an]other, especially when the document prepared in a particular font doesn't get downloaded in the other computer in the absence of the same font in the latter one" (2).

With the absence of Unicode standards corresponding to Nepali in Devanagari script, Nepali speakers who wanted to type in Devanagari script adapted and innovated variations of other scripts (e.g., Himali, Preeti, Kantipur, Sama). Every time a document was saved and reopened on a different computer, the script would be distorted, since "sometimes even a single page of document may contain several kinds of fonts which make[s] the downloading process even more complicated and time consuming" (Prajapati, Shrestha, and Jha 2008, 2). Furthermore, as Prajapati, Shrestha, and Jha (2008) also explain, "present keyboard layouts have been designed without much research on how they impact the typing behavior of the users. These layouts are irregular in terms of statistical distribution of the keys due to which the keyboard layout puts excessive and disproportionate stress on the fingers which [in the] long term can cause adverse effects" (1). In other words, the lack of inclusion of Nepali Devanagari script, coupled with the lack of usability testing on keyboards that incorporate that script, make it so that even contemporary Nepali speakers may opt to use Roman script to type content in Nepali, perhaps erasing a rich history of Nepali culture through the erasure of Nepali Devanagari script. Furthermore, the lack of usability research conducted with Devanagari keyboards makes these keyboards inaccessible for users with limited mobility.

At the same time, however, many Nepali speakers have developed innovative ways to maintain their learning of Nepali Devanagari script. For example, Nepali programmers developed apps like Typeshala, a Nepali typing tutor that teaches users how to convert Roman script into Nepali Devanagari (and vice versa). While the dominant form of technology (i.e., Unicode) may not have been initially available to users of Nepali Devanagari script, as is the case for many marginalized communities that are excluded from dominant technology development, many Nepali speakers developed innovative tools, both digital and material, to maintain their language and writing system. In this way, the lack of availability of Nepali Devanagari script, inasmuch as it may have influenced some erasure of the script in digital spaces, also fostered creativity and innovation among Nepali speakers who blended scripts and language practices in their typing.

In the workshops at SAFAR, several participants noted that they did not have Nepali Devanagari script keyboards on their computers, even though these keyboards are now available. Many contemporary Nepali speakers rely on Nepali Unicode converters to transform Nepali Roman script to Nepali Devanagari script, which is what I noticed students doing during our workshop sessions. For some students, it was important to include Nepali Devanagari script on their professional websites and other digital projects as a way to represent their heritage and language. However, many also admitted that they do not type in Nepali Devanagari script on a regular basis, opting to use Roman script in informal chat spaces. For some students, Nepali Devanagari script keyboards are available on their cell phones, but some choose not to use this script because they've gotten in the practice of typing Nepali using Roman characters. Thus, the use of Nepali Unicode Converters has evolved as a form of user localization (Sun 2012) related to typing in Nepali, since Nepali speakers adapt content across multiple platforms—including word processing software, Nepali Unicode Converter websites, and other sites—to localize their writing into a form that represents users' own preferences for communication as well as their technological preferences and literacies.

I recently had a follow-up conversation with one of the students in my workshop who is a member of the Rai community and who speaks Rai language, one of the more than 100 Indigenous languages spoken in Nepal. During this conversation, the participant explained that many of Nepal's Indigenous languages are endangered and are looked down upon by Nepali speakers. Indeed, as Bal Krishna Sharma and Prem Phyak (2017) explain, neoliberal ideologies in Nepal continue to perpetuate the erasure of Indigenous languages in preference for the standardized use of both Nepali and English. Yet as Sharma and Phyak (2017) also clarify, "Although the 2015 Constitution recognizes the multilingual, multicultural, and multiethnic identity of Nepal and mentions all of the languages spoken within its territory as 'national languages,' Indigenous languages have not been used as official languages until now. However, there is a growing Indigenous activism and pressure towards creating space for Indigenous languages in education, mass media, and other domains of language use" (234–35). This growing awareness and activism by Indigenous communities is also made evident in the development of online keyboards and scripts. For example, while the Nepali Devanagari script is still largely imperfect and evolving (Prajapati, Shrestha, and Jha 2008), Indigenous communities have taken it upon themselves to design, test, and update digital scripts for several Indigenous languages, including variants of the Rai language, Newari language, and Limbu language, among many others. For technology designers seeking to standardize and Westernize Nepali language practices, the use of Roman script to type in Nepali is yet another marker of modernization and progressivism, as users can toggle between Nepali and English in digital spaces while using the same script. However, for some Indigenous communities that want to preserve their language in online spaces, it's important to develop usable scripts in Indigenous languages, thus leading to more effective, usable keyboards in Indigenous languages than those available in the more widely spoken Nepali.

A simple Google search for a Rai language keyboard, for example, leads to several resources for typing Rai language in digital spaces (https://omniglot.com/writing/bantawa.htm). Similarly, a search for a Limbu language keyboard also yields several plug-in resources, many of which have been developed by members of the Limbu community to facilitate the use of Limbu language in online spaces (GooglePlay 2020). These are just two examples, as multiple digital tools have been developed by Indigenous language speakers in Nepal who consistently work to teach and preserve their languages in online spaces. While more updated resources for using Indigenous languages online are always needed, it's important to note the contrast between the development of online technologies for Indigenous languages and for Nepali more broadly. When Western communication practices, including English and Roman script, are privileged, non-Western languages like Nepali are negatively impacted. Furthermore, because Nepali itself is sometimes positioned as a dominant language used to erase Indigenous languages of Nepal, Indigenous language speakers in Nepal have added impetus to develop typing technologies to aid in language preservation and revitalization efforts.

The centering of dominant languages can sometimes, both intentionally and unintentionally, lead to the fostering of communicative practices that erase traditional knowledge systems in favor of those most privileged. This is the case with the issues around using Devanagari scripts in online spaces, and it can also be the case with defaulting to English as the main form of communication in a pedagogical space like the workshops I co-facilitated. Defaulting to English as a neutral language can erase the linguistic expertise in the room, even when students or participants in a particular context speak English fluently. Acknowledging the positionality of the languages used in a space and making space for collaborative translation and language negotiation can thus help technical communication researchers and teachers make more informed decisions about the power dynamics they are perpetuating and embracing, even with the best intentions. Language positionality is also an important yet often unrecognized element that impacts participation in collaborative research.

## PARTICIPATORY DESIGN WITH STUDENTS IN NEPAL

The discussion of keyboards and scripts available to Nepali users illustrates some of the possibilities community-driven participatory design can have in South Asian contexts, particularly in communities with such linguistic richness. For the purposes of this project, we (the workshop facilitators) initially chose to focus on participatory design as the framework for our workshop due to this method's potential for broad engagement. Rather than simply facilitating lectures or technology workshops that focused on learning a single tool at a time, we chose participatory design as a way to shape the workshop's structure along with, rather than just for, students. Since all the workshop participants were graduate students and/or professionals, we knew each participant would have their own particular interests, and we didn't want to impose a single project or focus for the workshops without our participants' input. However, as

I document in this section and throughout this book, making space for participation does not always guarantee that participants will feel comfortable sharing their perspectives and feedback.

Indeed, participatory design has long been described as a democratic way to engage in technology design, rooted in a Scandinavian tradition that "advocates for the full and direct participation of end users within the design process" (Rose and Cardinal 2018, 10). As Emma Rose and Alison Cardinal (2018) further note, participatory design methods emphasize the importance of "engaging users throughout the design process, giving them a seat at the [design] table and a full voice in the design process" (10). Yet when working in multilingual spaces, especially with multilingual communities of color in colonized contexts, participatory design methods may need to be fully adapted and localized to be effective and to prevent harm (Agboka 2013; Acharya 2018; Chavàn 2005; Rose et al. 2017; Sackey 2020). As Apala Lahiri Chavan (2005) explains, "All the methods [for design] used in the Western world are based on the premise that participants will find it easy to articulate their thoughts and feel comfortable to say what works for them and what does not" (1). This perspective on participation, however, does not consider cultural, racial, and linguistic dynamics at play in participatory design and is thus "heavily loaded in favour of certain cultures and against others" (Chavan 2005, 1). As Godwin Y. Agboka (2013) argues, in participatory research, including localization, "significant issues such as local knowledge systems, political issues, economic implications, and legal systems prevailing at users' sites during the localization process are overlooked, if not ignored" (29). The same is true for participatory design and user experience research methodologies that benefit Western-based technology designers while merely extracting knowledge from local communities (Cardinal, Gonzales, and Rose 2020; Costanza-Chock 2020).

In our participatory design workshops, we found that rather than assume a sort of "linguistic neutrality" in our workshop space by resorting to English communication, we should emphasize and leverage linguistic difference as a way to both build trust and establish community with participants, making translation a common collaborative activity we undertook as a group. Typical interpretation protocols, where one interpreter translates all information into a designated language, were not necessary in this space since everyone generally spoke English. However, rather than rely on English completely, we found it important to leverage the collective linguistic expertise of the room when engaging in conversations and design activities. Specifically, we found that participatory translation activities could be useful starting points for discussion, allowing participants to define and translate different concepts across languages and to then use differences in translation to discuss technology use and design. To really engage in participatory design, we as a group needed to start by (re)defining and thus translating participation for the particular context in which we were working rather than relying on Western-based assumptions about what participation entails.

*Participatory Translation*

As a group, one of the projects our workshop team decided to take on was creating professional websites for each participant, where each person could showcase their research interests, community work, and other aspects of their identities and professional goals. The team began designing these websites during the second week of the workshop, with each participant creating their own individual site as the whole team engaged in design conversations to see each other's approaches while also taking time to usability test each other's designs. During a workshop session, we introduced the concept of usability testing, providing several sample usability tests, including several scenarios representative users could be asked to complete in their assessment of a website. Workshop participants first completed scenarios that we wrote in English, using these scenarios to analyze a sample website. Then, participants were asked to write their own scenarios and to then conduct usability tests on their professional websites.

Because the sample scenarios we provided were written in plain language and seemed to ask users to complete simple tasks, I initially assumed that our participants/students would understand the instructions and complete the required tasks. However, once we started the usability testing activities, I realized immediately that the group was hesitant to engage in the activity, and there seemed to be some resistance to the notion of conducting "traditional" usability activities. Upon noticing that students were not completing the scenarios, I realized that there was some pushback against this activity that I wanted to be honest about and address with the group. While students seemed to understand the scenarios themselves, in the moment of participation, I noticed that they seemed reluctant to share their opinions and perspectives. Students commented that the websites they were testing were "good," without expanding on their perspectives or providing extensive feedback. This echoes other work showing that usability methods developed primarily from a Western perspective affect responses in usability evaluations, especially when participants and researchers are from non-Western cultures (Vatrapu and Pérez-Quiñones 2006; Winschiers and Fendler 2007). Participatory design methods need to be adjusted to respect local knowledge, norms, and communication styles (see Chavan 2005; Clemmensen 2011).

In this particular context, I decided to attempt to de-center or de-neutralize English as the lingua franca of our workshop space to foster conversation not only about the websites that were being reviewed but also about the entire concept of usability testing. In a traditional model for this usability study, the facilitators may have pushed students to engage further or provide more feedback about the product tested. The goal here would be to privilege traditional usability study methods in service of having students complete the activity and share their opinions. However, the goal of this particular activity was not for students to share their perspectives on a specific tool but rather to learn about the function of usability testing and to participate in the activity so they could later apply what they learned in testing their own websites. At this point, rather than ask students to continue completing the scenario instructions, I decided to input parts of a scenario related to a video on a website into Google Translate, translating from English to Nepali. For example, the task "Imagine you are learning about Nepal for the first time. What do you learn by watching this video" was translated to "कल्पना गर्नुहोस् कि तपाइँ पहिलो पटक नेपालको बारेमा सिक्दैहुनुहुन्छ । यो भिडियो हेरेर तपाइँले के सिक्नुहुन्छ?"

In that moment, I wanted to de-center my role as "expert" in this usability study by asking questions related to students' own languages and expertise. This type of reframing, I thought, might foster more participation from the room while also leading to conversations that could inspire students to later design their own usability tests. Having done previous research on the problems with digital translation tools, including Google Translate, I imagined that the translation of the usability study scenario would have several issues and that these issues might motivate students to provide some feedback on the (in)accuracies of Google Translate.

When they first saw the translation produced by Google Translate, students who spoke Nepali began to comment on the translation discrepancies and mistakes, and the whole class engaged in a discussion about what the scenario was really asking students to do. We were also able to expand on our discussion of Devanagari versus Roman script, where students explained that they could read Devanagari even when they did not feel comfortable typing in that script. We thus engaged in a discussion about the many ways digitization, including the use of digital translation software as well as word processing software, changes the representation of language and culture in both material and metaphorical ways.

We then began to input different scenarios into Google Translate, not only to see the translations but also to see how language is reshaped and sometimes warped through the introduction of a digital interface. Students then pointed out that this same type of reshaping and distortion might happen when our own identities are conveyed through digital platforms such as professional websites, particularly when we are designing these websites in English within the constraints of a specific content management system like Wordpress or Squarespace. How could we really design professional websites that represented students' interests and backgrounds? How do the constraints and affordances of a website shape or influence the things students may want to highlight about themselves for public audiences? These were the questions that emerged through a discussion of translation in the context of a usability study.

While the translation activity was initially incorporated into the discussion to foster participation in the user scenarios, opening up space for translation in this context also allowed our team to transform the entire usability activity into something that was localized to the interests of our group. Through this discussion, I realized that the issue with the initial scenario in English was not that students did not understand the instructions but rather that they were reluctant to openly critique a product, particularly because the format of a usability study did not successfully foster conversations in this context. After seeing the inaccurate translation of the scenario, students felt increasingly comfortable critiquing the translation and making translation recommendations. This type of engagement activity prompted students to open up and begin to feel more comfortable sharing their perspectives. Since we as workshop facilitators listened to students' translation critiques, the perception of the activity as a whole shifted from not wanting to critique a product to collectively voicing perspectives and opinions alongside the workshop facilitators. In this case, the translation activity shifted the dynamics in the room enough to encourage students to see that their feedback was truly welcomed. In addition, the introduction of translation activities during a conversation about usability studies helped the entire workshop team adapt and localize the methods for engagement and participation used in the workshop as a whole. Rather than trying to simplify user tasks into simple scenarios presented in plain language within the constraints of the English language, participants in these workshops actually preferred to complicate the nature of language by seeing various translation options as meaning is negotiated through digital interfaces. The act of translating collaboratively in this context allowed for participatory design to leverage the power of translation and to intentionally include participants' linguistic skills in the design process.

### *Affinity Diagramming*

Following this user scenario activity that took place relatively early in the workshop series, I continued using translation as a way to encourage participation, especially when students seemed reluctant to share their honest perspectives or opinions or when we just needed motivation to get the conversation going. For example, another common UX and participatory design activity is affinity diagramming or mapping, which is an activity used to condense a large amount of quantitative data into major themes or patterns based on participants' ideas (see https://methods.18f.gov/decide/affinity-mapping/). For this activity, facilitators usually give participants sticky notes and ask them to freely associate specific terms related to a design, and then a couple of selected participants (sometimes along with the facilitator[s]) group common themes and ideas together to help guide decisions about a design.

While affinity diagramming is often used to gather feedback about a design, during our workshops at SAFAR, the workshop facilitators found it useful to incorporate affinity diagramming as part of participatory translation activities. For example, at the beginning of our discussions on particular days, we would hand out sticky notes to all participants and ask them to provide translations and definitions of relevant terms into Nepali and/or English. We asked students to translate and define terms such as *participation, digital, design,* and other disciplinary concepts like *South Asian studies* and *digital humanities.* The idea here was to set off the discussion with an emphasis on the fact that these terms have various definitions and to emphasize the point that we as workshop facilitators (and me specifically as a non-Nepali) did not have set definitions or translations for these concepts. Once participants had a chance to define these concepts and provide translations individually, the whole group got up and started to make connections between sticky notes, finding pattern and discussing discrepancies in the ways each member of the group defined and translated each term. Figures 5.1, 5.2, and 5.3

*Figure 5.1. Shreejana defines "participatory."*

are images of participants discussing and grouping together definitions and translations of the term *participatory*.

Words and definitions associated with the concept of "participatory" included "democracy," "diversity," "representing," and "home knowledge." While these definitions may not be commonly associated with the concept of "participatory," these associations were made by participants at the workshop and were intended to represent how the group as a whole understood what it meant to participate. Unlike traditional affinity diagramming activities where the goal is to extract ideas and knowledge from participants for the benefit of a design (Cardinal, Gonzales, and Rose 2020), the goal of incorporating affinity diagramming into this workshop series used participants' expertise in translation to establish a shared language for describing the goal of the workshop and the pedagogical space the group was sharing. For this workshop, having participants learn the process of participatory design and user experience research was more important than landing on a clear set of design recommendations for a company or product. When participatory design was grounded in multilingualism and translation, participants were generally better able to see how their expertise can contribute to the conversation at hand and to the goal of the workshop more broadly.

Figure 5.2. Pragya defines "participatory."

Figure 5.3. Dipak and Laura define "participatory."

*Material Technologies*

While incorporating translation into UX activities encouraged more participation from students in our Kathmandu workshops, we also needed to develop other activities and methods to help the group generate ideas about the types of projects they would like to undertake during our time together. To start, it was important to counter assumptions about students' "technological literacy," since several students were initially concerned with their ability to successfully build websites and other platforms. Long-standing conversations about technological "literacy" have positioned South Asian communities, and Nepali communities specifically, as deficient in technological expertise; these false assumptions undoubtedly make their way into classroom and workshop spaces. For example, in discussing the role of UX in South Asian contexts, Md. Faruk Hossain (2020) claims that "Southeast Asian countries like Myanmar, Pakistan, Nepal, and Sri Lanka would benefit from more UX domain expertise as they are not as mature from a UX perspective as India" (n.p.). This reference to a lack of technological "maturity," in addition to the generalization about what encompasses "South Asia," is echoed in many conversations about Nepali communities that have historically been excluded from and oppressed in technology-driven research and practice (see Poudyal 2018).

In the case of the students at SAFAR, it was important to establish a workshop space where students could feel comfortable trying new digital tools without needing to perform expertise. After all, SAFAR is a space that fosters innovation and highlights the value of digital composing, particularly when bringing together multiple perspectives on South Asian studies. Thus, I wanted to embed opportunities for students to build and design things that really reflected their own interests rather than have students perform activities they thought the facilitators would approve of. Rather than spending time teaching individual tools, we as workshop facilitators provided an overview of several possible digital tools but also told students that they themselves could probably find other tools to use that would be more suitable to their own interests.

To facilitate conversations about technological interests in low-stakes environments, I incorporated conversations from research in technical communication and related fields that point to the value of culturally localized approaches to technological innovation (Banks 2006; Haas 2007, 2012; Sun 2012). For example, I used Angela Haas's (2007) positioning of "Wampum as Hypertext" to envision the types of material and digital technologies students at SAFAR already engaged with on a regular basis. As Haas (2007) explains, reaching "digital and visual rhetorical sovereignty" requires a revisioning of history to focus on how Indigenous communities have been innovating and sustaining material technologies long before technologies like hypertext were "discovered" in Western contexts (78). Through this historical revisioning, Haas

(2007) positions "American Indians as the first known skilled multimedia workers and intellectuals in the Americas," providing a model for how other digital rhetoric scholars can pay closer attention to the histories embedded in "new" and "emerging" technologies across the globe (78). Building on this work, we as workshop facilitators wanted to find ways to illustrate to students how their own cultures, languages, and backgrounds functioned as and sustained technologies that facilitated complex rhetorical work. We wanted to highlight students' innovative practices, language skills, and backgrounds as opportunities for technological innovation.

In one workshop session, I extended Haas's discussion of wampum as hypertext by beginning our lesson on digital innovation with an activity developed by Chicana feminist educator J. Estrella Torrez (Torrez et al. 2019). In this activity, called "the yarn activity," participants use a ball of yarn to draw connections across everyone in the room, throwing the yarn to create a web that connects all participants as they build community.

In the workshop, we began by sharing our research interests and what we wanted to build. After one person shared, all others who resonated with the initial person's comment raised their hands, and the initial sharer hung on to one piece of the yarn while throwing the ball of yarn to another participant, who then repeated this same process. At the end of the activity, all members of the group were holding on to a piece of yarn while the length of the yarn became intertwined among all participants, thus simulating hypertextual connections among different websites, different students, and different designers. While relatively simple and unrelated to digital technologies, this community-building activity helped foster participatory design in this context not only by encouraging students to share their ideas but also by helping all people in our shared space to build a collective community. In this way, as I explained in our debrief discussion, no one person needed to have all the technological expertise or all of the ideas. Instead, the collective goal of the group became to leverage everyone's skills and contributions to design digital tools and platforms that would benefit the group's interests while also countering ongoing false assumptions about technology use and representation in Nepal.

This material, embodied activity for illustrating connections was referenced frequently by students throughout the workshop as they began building their websites and expanding their projects; in fact, workshop participants recently used this activity in their own workshops and community-building spaces. In the end, each member of the group decided to build their own individual website but to theme their projects in relation to one of SAFAR's overall themes and academic projects: Rethinking South Asia.

## RETHINKING SOUTH ASIA

As one of SAFAR's primary missions is to "create a venue for dialogue and debate between international scholars working on South Asia" (http://www.safarsouthasia.org/mission/), one of the projects SAFAR researchers are constantly engaged in is that of "Rethinking South Asia"—developing various methodologies, comparative studies, and entry points to understand South Asia from multiple intertwined and interdependent perspectives. Pushing against long-standing infrastructures that position some South Asian contexts as central, the Rethinking South Asia project at SAFAR provides space for multiple researchers to engage with topics and areas of study that can illustrate the multiplicity of perspectives South Asian scholars across countries bring to the discipline.

During our workshop series, by constantly referencing our interconnectedness as illustrated by the yarn activity described in the previous section, our group had multiple conversations about what it could mean to leverage what we learned about technology design as the group continues to rethink South Asia from multiple perspectives. We continued using participatory design activities not only to translate content across languages but also to innovate ideas for what it means to "Rethink South Asia" in digital spaces, particularly through each student's perspective. By the end of the workshop, each participant drafted their own research statement as part of their individual website, where they articulated their goals as well as their orientations to South Asian studies. Rather than summarize students' dynamic projects, below, I include excerpts from selected students' research statements and bios to illustrate the breadth of the projects that continue to take shape beyond a single workshop.

## STUDENT BIOS AND REFLECTIONS

*Renuka Khatiwada*

My name is Renuka Khatiwada. I teach English composition to the undergraduates in Oscar College of Film Studies, Kathmandu, Nepal. I am a photographer-researcher for the South Asian Foundation for Academic Research (SAFAR). I am working on visual studies projects. My exhibitions were on *My Female Deities in Urban Spaces* in Oct. 25–26, 2018, in Kathmandu; *Our Earth, Our Architecture, My Photography* in Jan.

15, 2019, Dhaka; and *Chittagong Symphony* in Oct. 15–17, 2019, Chittagong, Bangladesh. As a videographer, I have worked on two documentary projects, *Housewife, Renouncer, and Artist* and *Four Gray Walls, Four Gray Towers.* My recent research on "Women in the Streets" documents the experiences of women in the marginalized communities in the cities of Nepal and Bangladesh. I archive these projects on a digital humanities platform.

I participated in the Critical Digital Humanities workshop in June 2019 held in Kathmandu, Nepal. The program was organized by SAFAR and sponsored by the United States Educational Foundation in Nepal (USEF). Dr. Laura Gonzales's supervision helped us understand technology in multiple new ways. The most significant aspect of the workshop is the significance of humanistic learning in the digital platform with the idea that to be a scholar is to comprehend information and knowledge technologically, which paves the way for international and rapid academic connectivities. I also understood that conventional classroom situation is not sufficient to be contemporary in modern educational system.

*Prateet Baskota*

I am Prateet Baskota. I am currently a MPhil scholar in Kathmandu University School of Education, Nepal. Besides that, I am a Spanish language instructor in Daffodil Boarding and Public School. Additionally, I was once an IELTS, American GED trainer, and English language teacher. I am interested in digitalization and education so I did a few presentations on "Boosting Fresh Mode of Teaching English Story by Using Animation," "Discourse in Eccentricity Ex-Centricity through History of English Education," "Collaborative Learning and Generating Ideas!" "Maus: Betrayal, Holocaust, and the Graphic Fiction: An Independent Paper," and "Use of Internet in English Classroom." I am also a researcher in Kathmandu University as a student, which means I learn by doing. After completion of an MA in English from the Institute of Advanced Communication, Education, and Research (IACER), I am engaged with different research centers in academia.

*Pragya Dahal*

I am Pragya, an enthusiast in research, program implementation, and management and advocacy through communication and media outreach. I am pursuing my MA/MPhil and have contributed as a guest lecturer at the Arunima Education Foundation. Recently, I worked as an advocacy and communication officer in one of the development organizations advocating for family forest farmers. My interest lies in the issues related to gender and social inclusion, environmental conservation, which perfectly matches with the experience I had gained from my professional career as a development worker. As a person, I am fond of advocating for the rights of people (esp. gender-based violence) and women's rights. I further want to develop my career in environmental conservation and family forestry. I am also a good communicator and presenter and I use my skills for sensitizing people about their rights. Personally, my keen interest lies in volunteering, building up networks, exploring new places, and writing.

*Sedunath Dhakal*

My name is Sedunath Dhakal. I have degrees in English (MA and MPhil). My areas of interest are culture, identity, conflict studies, as well as literary theory and philosophy. I teach critical thinking and practical criticism, essay and short stories, and managerial communication at Koteshwor Multiple Campus, Koteshwor, Kathmandu. I have been working on cultural and heritage conservation by digitally archiving. My recent research work is *Borderland, Culture and Contestation: A Case from Nepal-India Border.*

*Nandita Banerjee*

My name is Nandita Banerjee. I am an independent researcher and writer who has lot of interests in film and literature. I am an assistant professor in Katwa College, University of Burdwan, and guest lecturer in the Department of History, Rabindra Bharati University.

Specializations: Economic history of India, role and representation of women in popular Hindi cinema in the post-independent era. I completed a master's and PhD from Visva Bharati Santiniketan. My dissertation is on "The Agrarian Society of Central Province 1818–1883." Present areas of interest: popular cinema, gender, and sexuality. I am planning to write a book on the portrayal of women in Hindi cinema in the post-independent era. I am attending one workshop on digital archiving in humanities in Nepal, and I have plans to make a short film.

*Dipak Bastakoti*

मेरो नाम दिपक बस्ताकोटी हो । ब्यबसायिक रुपमा नेपालको हिमालयमा पदयात्रा पेशामा संलग्न रहेको दुई दशक बढी भयो । मैले आफ्नो पेशा संगै विश्वविद्यालयको पढाईलाई अघि बढाउँदै सन् २०१७ मा पोखरा विश्वविद्यालयबाट अंग्रेजीमा एम.फिल.डिग्री प्राप्त गरैं ।

My name is Dipak Bastakoti, and I have been professionally engaged as a trekking leader in the Nepal Himalayas for more than two decades. Along with my profession in the mountains, I continued my university studies and obtained a MPhil degree in English from Pokhara University in 2017.

मेरो पेशाको बखत मैले भिन्न सामाजिक, आर्थिक, वर्ण, समुदायका विभिन्न देशहरुबाट आउने व्यक्तिहरुसंग अन्तरक्रिया गर्ने, र फरक-फरक संस्कृति, रहनसहन, मुल्यमन्यताहरु बारे मौका पाए । बाल्यकालमा मैले सिकेको हामी , सामुहिकता, समुदायिकता जस्ता मान्यताहरुको सट्टा म , 'व्यक्तिगतता', र प्रतिस्पर्धा' जस्ता मान्यताहरु हावि भएको जस्तो लाग्न थाल्यो । सगरमाथा क्षेत्रको पदयात्रामा गएका बेला सुनेका, केही देखेका, र केही हिमाल आरोहण सम्बन्धि पुस्तकहरु पढेका आधार मा हिमाल आरोहणको इतिहास र शेर्पाहरुको जीबन औपनिवेशिक तथा साम्राज्यबादी प्रभावमा रहेको नमिठो अनुभाव भयो । वर्तमान समेत यस क्षेत्रमा स्वार्थकेन्द्रित व्यक्तिबादी पुँजिबादी प्रतिस्पर्धा रहे को र यसैको चपेटामा बर्षेनी धेरै शेर्पाहरुको अमुल्य जीबन खेर गएकोले मलाई बिचलित बनाउन थाल्यो ।

In my work, I have the opportunity to interact with people from varied social, economic, racial, and community backgrounds, and I understand various different values and cultures. Values that I was accustomed in my childhood, such as community, the notion of "we," and collectivity, are frequently somehow challenged with individualism, the notion of "I," and competition. What I have heard [and] seen during trekking trips in the Everest region, and know through reading various books on mountain literature is something I came to realize dismally—that the history of climbing and the life of Sherpas have been dominated by colonial and imperial impacts. I became further downhearted from the contemporary capitalist competition in the mountain that claims scores of precious lives in the Himalayas.

एम. फिल. शोधपत्रको लागी शेर्पाको जीबनसंग सम्बन्धित रहेर केही उत्तर-औपनिवेसिक, साम्राज्यबादी र महिलाबादी कोणबाट अध्ययन गर्ने सोचेको थिएँ तर केही प्राबिधिक कठिनाईको कारण यस्तो हुन सकेन । म यस्तै धडधडी बोकेर "क्रिटिकल डिजिटल हुम्यानिटिज एण्ड पार्टिसिपेटरी डिजाईन" भन्ने कार्यशालामा भाग लिन पुगेको थिएँ । यो कार्यशाला हुन भन्दा पहिले सन् २०१७ मा मैले अध्ययन गरेको संस्था आइएसिइआरमा (Institute of Advanced Communication, Education, and Research—IACER) मा संचालन भएको #DHNepal2017 seminar नामको गोष्ठीको समापन प्रवचन सुन्न गएको थिए । क्यालिफोर्निया स्टेट विश्वविद्यालय, नर्थरिज का डिजिटल हुम्यानिटिज केन्द्रका निर्देशक, प्राध्यापक स्कट क्लेइम्यानले किन मानविकीका विविध आयामहरुका सामाग्रीहरुको संकलन, अभिलेखिकरण र डिजिटलाइज गर्न आबस्यक छ भन्ने बिषयमा उल्लेख गर्नुभएको थियो । यस बिषयमा मेरो अभिरुचि बढेकोले उत्साहका साथ कार्यशालाबारे सुचना पाउने बित्तिकै सहभागी हुन नाम दर्ता गराएको थिए ।

For my MPhil dissertation, I desired to work on the life of Sherpas from the vantage point of postcolonialism, imperialism, and feminism.

But it was unsuccessful due to some technical reasons. With this discouragement I participated in the mentioned workshop, Critical Digital Humanities and Participatory Design: A Workshop Series in Nepal. Prior to this workshop, I attended the closing lecture event of Professor Scott Kleinman, director of the Center for Digital Humanities, of California State University Northridge at the end of #DHNepal2017 seminar at IACER (Institute of Advanced Communication, Education, and Research) in 2017. His remarks on why various resources of humanities should be documented, archived, and digitized impressed me. My interest escalated on the topic; thus, I signed up for the #DHNepal2019 workshop as soon as I learned about it.

यस कार्यशालामा सह-संचालक भएर आउनुभएको–टेक्सास विश्वविद्यालय, एल पासोमा विद्यावारिधि गर्दै रहनुभएको विभूषणा पौड्यालसंग पहिल्यै चिनजान रहेको, तथा सफर (SAFAR—South Asian Foundation for Academic Research), कास्सा (CASSA—Center for Advanced Studies in South Asia) का बिभिन्न परियोजनाहरुमा सहकार्य गरेकोले म उत्साहित र केही रोमाञ्चित थिएँ। त्यसमाथि कार्यशालाको मुख्य संचालक अमेरिकी विश्वविद्यालयको प्राध्यापक। मलाई थप उत्साहित गराउनमा औपनिवेशिक विरासतले नै छाडेको जबरजस्त छाप ममा पनि छँदै थियो कि पश्चिमाले दिएको जस्तोसुकै ज्ञान पनि मुल्यवान हुन्छ भन्ने।

हामिमा एउटा भावना र सोच छ कि अमेरिकी शैक्षिक संस्थाहरु र प्राध्यापकहरुले एक किसिमको साम्राज्यबादी र स्वेत-उच्चता प्रभुत्वको प्रतितिधित्व गर्दछन्। यो वास्तवमै देखिन्छ पनि। मैले सन् २०१९ मा एक अमेरिकी विश्वविद्यालयमा निकै राम्रो बृत्तिमा विद्यावारिधि गर्ने अवसर प्राप्त गरेको थिए। जब मैले अमेरिका प्रवेश अनुमतिको लागि आवेदन दिए, नेपालस्थित अमेरिकी दुतावासले मेरो आबेदन कैयौं पटक अस्विकार गऱ्यो। मैले प्रबेशाज्ञाको लागी लिइने अन्तर्वार्ताका क्रममा अधिकारीहरुसंग साम्राज्यबादी शक्ति र स्वेत-उच्चताको घमण्ड अत्यधिक रहेको अनुभब गरें। हरेक पटक यि अधिकारीहरुले अपमानित गरेको मैले महशुस गरें। साम्रज्यबादी प्रभुत्व र घमण्ड तथा औपनिवेश विरुद्धको हाम्रो लडााईको बारेमा म तल आउनेछ।

I was excited and thrilled to sit in the workshop, since the co-leader of the workshop was Bibhushana Poudyal, who was familiar to me during our various projects organized by SAFAR and CASSA. On top of that, the leader of the workshop was a professor from an American institution. There is a general understanding in us carried down from the colonial era that anything the West imparts is the most valuable. So, I was further thrilled to participate the workshop.

There is a feeling in us that American academic institutions and professors reflect a site of white domination, an imperialist impression. For example, I requested a visa to study at a US university, where I was awarded a very impressive scholarship in 2019, multiple times, but the visa officers at the embassy in Nepal exhibited imperialist power, complex of white superiority impersonation, and treated me with humiliation each time. I was consistently denied this visa. I will come to the topic of the colonial, imperialist struggle we are fighting later.

कार्यशालामा प्राध्यापक लाउराले माथी उल्लेख गरे जस्तै हामी विविध क्षेत्र, अनुभव, बिज्ञता, र अध्ययन चासो भएका सहभागीहरु थियौं । कार्यशाला र हामी सबैको उद्धेश्य आ-आफ्नो अध्ययन-अनुसन्धान चासोलाई कसरी प्रविधिसंग र डिजिटल संसारसंग जोड्न सकिन्छ भन्ने जान्न उत्सुकहरु थियौं ।

In the workshop, participants were from diverse areas and had various expertise and research interests, as Professor Laura mentioned earlier in this chapter. The initial goal of the workshop was to learn about how we can integrate our research interests and areas of study with the digital world.

कार्यशालामा हामिले उपनिवेश र साम्राज्यबादले दक्षिण एसिया समग्रमा र बिशेषगरी नेपाली भाषामा पारेको कुप्रभाव बारे छलफल गर्‍यौं किनकी यि भाषाहरुको कम्प्युटर र डिजिटल संसारमा प्रयोग गर्नुमा अत्यन्तै कठिनाइको महसुस हामि सबैले गरेका छौं । संगसंगै हाल अगाडी बढिरहेको उत्तर-औपनिबेशिक आन्दोलनहरु र यसको अरु उचाईको आबस्यकताबारे पनि घनिभुत बिमर्श भयो ।

प्राध्यापक लाउराले माथि उल्लेख गरे जस्तै नेपालमा अंग्रेजी भाषाको ज्ञानलाई प्रतिष्ठाको रुपमा, सम्भ्रान्तले जान्ने भाषाको रुपमा लिने समस्या एकातिर छ भने अर्कोतिर कम्प्युटरमा लेख्न ढिलो र अफ्टेरो हुने तथा भिन्न कम्प्युटरमा लेखेको नखुल्ने वा पढ्न नमिल्ने जस्ता प्राविधिक समस्याहरु रहेका छन् । स्वर र व्यञ्जन संगै धेरै संयुक्त व्यञ्जन बर्णहरु भएको नेपाली भाषाको निम्ति अंग्रेजी भाषाको मानकमा बनेको कीबोर्ड अफटेरो छ । नेपालीमा लेख्न मिल्ने गरी किबोर्ड स्थापना गर्न त मिल्छ तर एउटै एकैचोटी नेपाली र अंग्रेजीमा लेख्न निकै कठीन पर्दछ । यही लेखको लागी पनि मैले पटक पटक चित्रमा देखाएको जस्तो प्रतिकहरुको सहारा लिन पर्‍यो । कैयन अंग्रेजी शब्दहरुको उपयुक्त नेपाली अनुवाद फेला पार्न सकिंदैन एकातिर भने धेरै शब्दहरुको नेपाली रुपान्तरण प्रयोग गर्नं झन आमरुपमा बुझ्न कठीन बनाउन हुनेछ (यस्ता अंग्रजी शब्दहरुलाई मैले इटालिक्स मा राखेको छ) ।

Our discussion in the workshop was on the colonial and imperialist repercussions that impact the languages of Nepal in particular and South Asia in general, since the use of these languages in digital spaces on the computer is an arduous task. We further discussed vigorously various postcolonial movements and their impacts on language and technology.

As Professor Laura has mentioned in this chapter, English in Nepal has been considered indispensable to show superiority in society, [English is] the language of elites on one hand and on the other hand it is onerous to write, as it malfunctions on different computers. The keyboard prepared for the English-speaking world is insufficient for the Nepali language, where there are many vowels, consonants, and complex vowel combinations. Keyboards can be installed to write in English *or* in Nepali, but it is still very difficult to type in both languages within the same document. Even writing this piece, I took an extremely long time and had to insert special symbols. Furthermore, even after trying hard, I could not find many translations in Nepali for many English words and in some cases an English word is better to use [than the] Nepali translation (I write those English words in Nepali script and put them in *italics*).

बह्य मात्र नभई प्राध्यापक लाउराले माथि बताएजस्तो नेपालभित्रै पनि अन्दाजी १०० जति मातृभाषाहरु नेपाली भाषाको लगभग औपनिवेशिक भन्न मिल्ने प्रभावमा परेर समस्याग्रस्त छन्। यस्तै गरी, पश्चिमा देशहरुमा भएको प्रविधि र विज्ञान क्षेत्रको बिकास हुँदा जन्मिएका कैयन शब्दहरु नेपाली भाषामा उपलब्ध छैनन्। यस्तोमा हामि दुई खाले चेपवामा परेका छौं। एकातर्फ हामी अंग्रेजीमार्फत हाम्रा अभिब्यक्तिहरु स्पष्ट राख्न सक्दैनौं भने अर्कोतर्फ समकालिन विश्वका धेरै शब्दहरु हाम्रो पहुच भन्दा बाहिर छन्। यसैकारण, भाषामाथि भैरहेको औपनिवेशिक र साम्राज्यबादी दमन/प्रहार होस् वा मैले माथि उल्लेख गरेजस्तो दुतावास भित्र र बाहिर प्रकट भैरहेको स्वेत-उच्चता र घमण्डको बिरुद्ध होस् निरन्तर संघर्ष र प्रतिरोध जरुरी छ।

Within Nepal as well, there are more than 100 languages that suffered from internal colonialism through the oppression of the Nepali language as Professor Laura has mentioned. Likewise, even today, many words from science and technology that were invented in the Western world remain unavailable in our language. We are suffering from two-fold troubles. First, we are incapable of expressing our thoughts completely in English on one hand, and on the other hand we lack Nepali vocabularies for many terms of the contemporary world. So, there is a persistent need to fight against all kinds of colonial and imperialist suppressions: be it in language or in the behavior of the white officers at the embassy, as I mentioned above.

म अब मैले उक्त कार्यशालामा आफ्नो रुचिको परियोजनाको रुपमा सहभागीहरु र संचालकमा प्रस्तुत गरेको परियोजना बारेमा संक्षिप्तमा राख्न अनुमाती चाहन्छ।

Now I want permission to briefly reflect on my particular project that I outlined during the workshop.

सगरमाथा सहित हिमालयका सबै स्थानमा काम गर्ने शेर्पाहरुको जीबन, हिमाल आरोहणको अर्थतन्त्र, आरोहण बखत हुने राजनीति, र इतिहासको औपनिवेशिक तथा बर्तमानको पुँजिबादी प्रतिस्पर्धा जस्ता केही अवयवका वारेमा केही जानकारी प्राप्त गरेपश्चात मलाई भाभुक बनायो।

I was overwhelmed by the life and fates of the Sherpa community, the economy of mountaineering, the politics of climbing, and the capitalist competition in contemporary time, as well as the colonial competition in mountaineering enterprises historically.

नेपालको पर्यटन मन्त्रालयले सन् २०१४ मा प्रकाशित गरेको सगरमाथामा कायम भएका बिभिन्न विश्व रेकर्ड हरु बारेको पुस्तकमा सन् १९२२ मा ७ शेर्पा आरोहिको हिमपहिरोमा परी भएको मृत्य सगरमाथामा पहिलो अभिलेखित मृत्य भनेर लेखेको पाए, अरु जानकारी बिना नै (पेज ४)। सन् २०१२

सेप्टेम्बरको मनसलु दुर्घटना, सन् २०१४ को सगरमाथा खुम्बु आइसफल दुर्घटना, र सन् २०१५ को गोरखा भुकम्प पश्चात सगरमाथा आधारशिविरमा भएको घटनाजस्ता हृदयबिदारक अवसरहरुमा म सगरमाथा क्षेत्रमा उपस्थित थिए । यस्ता क्षणहरुमा आरोहणमा गएका शेर्पा य'वाहरुको नजिकका परिवारका सदस्यहरुको चित्कार, रोदन, र निराशा नजिकबाट हेर्न/देख्न पाए । यसपछि जब मैले केही आरोहण सम्वन्धित पुस्तकहरुको अध्ययन गरें, अधिकांशमा शेर्पाहरुको योगदान प्राय गाएब पारिएको वा उल्लेख गरे पनि अनुपयुकन तरिकाले प्रतिबिम्बित गराएको महसुस गरें ।

हिमाल आरोहणको क्रममा अरु आरोहीहरुको स्वार्थमा सेवा दिदादिदै मृत्युवरण गर्ने शेर्पाहरुलाई अंकमा गनेको र उनिहरुको परिवारको दर्द र पीडालाई नजरअन्दाज गरेकोमा मलाई छटपटि भयो ।

The Ministry of Culture, Tourism, and Civil Aviation of Nepal published a book of world records in Mt. Everest in 2020. In the book, the death of seven Sherpas in 1922 is mentioned as the first recorded death in Everest (4), without much explanation of them. I have been in the Everest region and have closely seen, [and] felt the cries, frustrations, and burdens of family members of Sherpa climbers in the 2012 Manaslu disaster, 2014 Everest disaster, and 2015 Everest Base Camp disaster after [the] Gorkha earthquake. And I was unable to find proper representation and contributions of the Sherpa community in the various books written by Western climbers.

Those Sherpas who died in the service of others are counted in numbers, and the troubles and pain left behind [for] their families are ignored. This in fact keeps me restless.

एसैले, मैले हिमालयमा समाधिस्त भएको शेर्पाहरुको अमुल्य जीबन र उनिहरुका परिवारजनले बगाएको आशाको उचित अभिलेख राखन उपयुकत हुने देखेर यो परियोजनाको प्रस्ताव उक्त कार्यशालामा राखें । यस परियोजना मार्फत मैले हिमालयमा अर्पित शेर्पाहरुको मानबीय, उत्तर-औपनिवेशिक, र महिलावादी दृष्टिकोणबाट अभिलेखिकरण गर्ने प्रयास गर्नेछ । म यस परियोजनामा सबै शुभेक्षक ब्यक्ति तथा संस्थाको मुल्यवान सहयोग र सहभागिताको अपेक्ष ागर्दछ' ।

Thus, I decided to work on collecting the tears and archiving the value of life of those deceased Sherpas. This project aims to archive all Sherpas who lost lives in the mountain from humanist, feminist, and postcolonial approaches. And I expect active participation and invaluable contribution of whatever kind feasible for the project from all the participants, well-wish[ing] individuals, and associations.

*Shreejana Ghimire*

As we see in this chapter, SAFAR has created a huge platform for different South Asian scholars to share their ideas and develop academic research through different methodologies, digital platforms, and technologies. SAFAR promotes and focuses on the collaborative development of digital technologies instead of relying on old systematic practices of research in South Asia. It is an open venue for the exchange of ideas and debates between international scholars working on South Asia, as well [as] in other places.

In 2019, I had the opportunity to participate in the three-week workshop series described in this chapter. The whole workshop series was focused on participatory design and user experience. It gave us new ideas to use the digital technologies in our specific research interest. This was my first time experiencing being a participant in a SAFAR workshop with different scholars from different fields and alongside international experts.

My expectation before going to the workshop was completely different, as I thought we would spend time doing traditional research and learning about the publication of articles and journals. After participating in the workshop, I realized the importance of engaging in research using advanced digital technologies. For example, I created a website where I blogged about gender representations and stereotypes in Nepal. The goal was to expand conversations about gender in Nepal and to have this site led by a female Nepali voice. This project changed my whole perspective and my expectations too.

After completing the workshop there were also some questions raised in my mind about the education system of South Asia, especially about the Nepalese education system. We are still sometimes behind in using and learning the latest digital techniques for research, not because we don't have internet facilities or due to a lack of media but because technology is not something that is emphasized in our daily educational environments. So I am left with questions, even after the workshop, such as: How can someone find relevant study materials for research that incorporate digital technologies? How can you make your research more advanced by using new digital media? What might be new possibilities in our research considering the use and development of technologies in multiple languages?

Workshops like this one will create the platform to interact and share ideas and share the new trends for better research. Collaboration between different scholars and experts will make our research and technologies better, and this workshop motivated me to see things differently and creatively. Looking forward [to] new workshops and new experiences.

*Shankar Paudel*

साउथ एसअिन फाउन्डेसन फर एकेडेमकि रसिर्च (सफर) ले सन् २०११ मा प्रयोगकर्ता अनुभव तथा सहभागी ढाँचा सम्बन्धी कार्यशाला सञ्चालन गरेको थियो। यो कार्यशालाको डजिटिल

प्रवधिकिो ढाँचा सकिाइ मेरा लागि एउटा ठुलो कोशेढुङ्गा साबति भयो। एउटा सहभागीका रुपमा म यस कार्यशालामा भाग लनि पाएकोमा सधैं गौरवान्वति महसुस गछु। आफ्नो कार्यलयको कार्यब्टस्तताता बावजुद पनि आधा दनिको बिदा मिलाएर लगतार तनि हफ्ताको लामो यस कार्यक्रममा म कुनै दिन पनि अनुपस्थति रहनिं ।

पहलिे पहलिे म प्रबिधीलाई एउटा समय बिताउने माध्यमको रुपमा मात्र लिने गर्दथे । आफुलाई प्रबिधिको अब्बल ज्ञान भएको प्रयोगकर्ताको रुपमा लिने गर्थे र मसँग ज्ञान नभएको होइन तर यो कार्यशालाले मलाई ब्यक्तिगत तथा अध्ययन-अनुसन्धानमा प्रबिधिको प्रयोगको एउटा फराकलिो दायरा देखाइदयिो, प्रबिधीलाई हेर्ने मेरो दृष्टिकोणमा परिवर्तन गर्दियो। यस अध्यायमा वर्णन गरिएजस्तै वेबसाईट बनाउने, ब्यक्तिगत परियोजनहरूमा काम गर्ने तथा अन्य डजिटिल प्ल्याटफर्मको प्रयोग गर्न सकिने ज्ञानले वास्तवमै मेरो ज्ञानको भण्डार बढाइदयिो। त्यसैगरि, यस कार्यशालका बहुआयामकि पक्षहरूले मेरो डजिटिल ह्युम्यानटिजि प्रतिको लगाव झनझन गहरिो भएर गयो। यस कार्यशाला पश्चात् म अहलिे एउटा राम्रो डजिाइनर भएको छु। सफरले आयोजना गर्ने धेरैजसो कार्यक्रमका ब्यानर, फ्लाइयर, पोष्टर तथा कार्डहरु मैले आफैं डजिाइन गर्नसक्ने भएको छु। मैले क्यान्भा, क्रेल्लोजस्ता सजलिै चलाऊन सकनि डजिटिल प्ल्याटफर्महरु को प्रयोग गर्ने गरेको छु। कार्यशालामा भएका कृयाकलापहरले हामी सबै एउटै समुदयका हौं र एक अर्कामा कुनै न कुनै रुपमा अन्तरसम्बन्धित छौं पाठ पनि सिकायो।

त्यसैगरी, कार्यशाला पश्चात् ममा एक कसिमिको आत्मविश्वासच्च्य पलायो जसले मलाई यसै सम्बन्धी केही अन्तर्राष्ट्रिय कार्यक्रमहरुमा भाग लनिे साहस जुट्यो। सबैभन्दा महत्वपूर्ण त यही कार्यशालाले मलाई एउटा डजिटिल ह्युम्यानटिजिको प्रशिक्षक पनि बनायो। म लगायत मेरा अन्य दुई मित्रहरु भएर हामीले सन् २०२० मा बाङ्लादेशमा ग्राजुएट तहका विद्यार्थीहरूलाई दुई दनिे डजिटिल ह्युम्यानटिजिको कार्यशाला सञ्चालन गर्न सफल भयौं । यसले मेरो उर्जा र आत्मविश्वास झन् बढायो। अझै भन्नुपर्दा, यसले म लगायत मेरा साथीहरुको रथिङ्किङ् साउथ एसयिा सम्बन्धी अनुसन्धानलाई ठुलो सहयोग पुगेको छ ।

त्यस कार्यशाला पश्चात् पनि निरन्तर रुपमा डजिटिल ह्युम्यानटिजि तथा अन्य अनुसन्धान सम्बन्धी सल्लाह र सुझाव दनिुहुने डा. लौरा गोन्जालेसलाई मेरो तर्फबाट धेरै धेरै धन्यवाद दिन चाहन्छु ।

शंकर पौडेल
काठमाडौं, नेपाल[2]

2. The UX and Participatory Design Workshop organized by [the] South Asian Foundation for Academic Research (SAFAR) in 2019 was a great milestone for me in learning about the design of digital technologies. As one of the participants of the workshop I always feel proud that I managed it in spite of my strict working hours. I took half-day leave daily for three weeks [so as to] not miss the workshop and I was always present.

I used to take digital space more as a medium for passing time. I used to think of myself as a techno savvy individual and I was good, too, but this workshop showed me a wide variation of using digital means in my academics and personal life too. But the workshop entirely changed my perspective about it. As this chapter demonstrates, the making of websites, working on individual projects, and knowing about multiple digital platforms truly have enhanced my knowledge of various technologies. In addition, the multiple facets of the workshop created a desire [to] learn more and now digital humanities [DH] and participatory design have become a passion for me. I turned into a designer after the workshop. I have been designing the flyers, banners, posters, and cards of the events and programs organized by SAFAR. I use the platforms which

## IMPLICATIONS FOR PARTICIPATORY DESIGN

As evidenced in the bios and narratives above, SAFAR students came into the workshop with a wide range of disciplinary interests, including environmental studies, international relations, literature, gender and women's studies, and health. Based on these interests and throughout our discussions related to participatory design, students extended their individual research statements and websites into prototypes for future design projects.

All of the project examples described are radically different from each other and represent just a snapshot of the extensive work these students continue to do beyond a single workshop. An important takeaway from a participatory design perspective was that to get to a place where students could develop their projects, the group as a whole needed to establish a space where innovation could happen. None of the students came into the workshop knowing exactly what they would create, and I certainly did not plan for all of these projects to emerge. However, through conversations and community-building activities, the group was able to decide on our common theme and to then find ways to, as participants said, "thread the yarn" across our interests. The reflections shared in this chapter were written by participants almost two years after the conclusion of the initial workshop. Students' enthusiasm and continued interest in UX and participatory design continues to shine through their reflections, and I believe this interest continues both because of students' commitment to the work and due to the relationships and space for discussion that were established through participatory translation and through an open awareness of the role language and positionality play in participatory design.

In chapter 1, I mentioned that I structured this book around three major areas of professional translation and interpretation: health, community, and law. The work presented in this chapter encompasses many elements of community translation, both because we used participatory design, and translation activities specifically, to build community with students and because the students reflected in this chapter are part of various

---

are free and could be used beautifully. The activities we participated in taught that we all are in a community and we are interdependent, in [one] way or another. It supports collaborati[on] and making a learning community.

Apart from this, I built up confidence after the workshop and have participated in multiple international programs related to it. Most important, the ideas I got in the workshop made me a trainer of basic DH in Bangladesh. In 2020, I, along with my two colleagues, conducted a workshop in Bangladesh for two days to graduate level students. It added confidence in me. Moreover, it has also supported me and our team in the research works of [the] Rethinking South Asia Project.

Nepali communities in schools, nonprofit organizations, and businesses. One of the key tenets of community-based translation is the reliance on community relationships to establish and sustain trust. This trust is particularly important when doing technology-related work with/in communities that are consistently positioned as technologically deficient, despite long-standing evidence about their intricate technological innovation (as evidenced in participants' navigation of various Devangari scripts).

In the context of the workshops at SAFAR, approaching community translation through intersectional, interdependent methodologies required an attunement to the history of English, Nepali, and various Indigenous languages in Nepal as well as the role colonization plays in contemporary classroom spaces and definitions of technological literacy. Furthermore, the emphasis on translation in participatory design, established in part through affinity diagramming and the yarn activity, reflected the interdependent nature of participation in this particular context. Rather than rely on a language interpreter to translate information among participants and facilitators, an interdependent approach to translation required the group to come together and establish what Rachel Bloom-Pojar (2018) refers to as a translation space, or the possibility for language negotiation in community, where the emphasis is not on reaching an "accurate" translation but rather on collectively making meaning with other members of a group. This emphasis on collaboration can help ease anxieties about being "good enough" with technology to make a contribution to conversations about technology design.

To be sure, the projects described in this chapter are imperfect and still emerging, and there are multiple other factors at play that will influence how the work of students at SAFAR continues to develop. Since leaving Kathmandu, I've had the pleasure of staying in touch with many students from the workshops, continuing to do my best to support the ongoing innovation of this research center and the students who make it successful. As SAFAR students taught me, the project Rethinking South Asia is recursive and never ending, which is also how I would describe successful participatory design in international contexts—as a recursive, never-ending process grounded in community relationships.

## CONCLUSION

While participatory translation provided one avenue for engaging workshop participants in discussion about specific concepts and ideas, the notion that participatory design practices and methods were going to inherently "welcome everyone to the [design] table" is inaccurate. Even with our adaptations to these participatory design activities, it was clear that for the group to truly innovate together, we needed to rebuild the entire notion of participation to begin with (Agboka 2013; Costanza-Chock 2020). As Agboka (2013) emphasizes, "We need a new paradigm of localization [and, I argue, participation] theories and methodologies that require approaches to culture that are not only sensitive to multiple contexts but provide perspectives of localization that draw inspiration from social justice theories" (29). While Agboka (2013) rightfully argues that intralingual factors are not the only elements that should be considered in social justice–driven design, the multilingual experiences presented in this chapter illustrate how language and translation can be leveraged as entryways into the deeper infrastructures that govern contemporary international research. To engage social justice approaches to participatory design, it's important to recognize the role language positionality plays in any collaborative space, particularly in non-Western contexts, as well as in multiple contexts within non-Western settings. Throughout this process, it's important to not make generalizations about how language and technology function in a particular space and to take time to learn about participants' histories and backgrounds and how these factors always impact any research setting. Language positionality and translation activities can help illuminate how Western, English-dominant frameworks can shape not only contemporary communication practices in global education but also the technologies through which this communication happens in digital spaces. At the same time, fostering relationships with research participants and researching the language dynamics of any research context can provide frameworks for understanding the ongoing innovation communities engage in as they work to sustain and extend their linguistic and cultural practices across space and time.

As many scholars of race and technology note, the algorithms of oppression (to use Safiya Umoja Noble's [2018] term) embedded within the designs of existing technologies are so rife with discrimination that they cannot be retrofitted (without a complete re-design and rebuild) to adequately support the activities and needs of non-normative bodies and non-white/Western/English-dominant audiences. In the case of the technologies built and used by participants at SAFAR, there is a constant balance among the innovation students engage in based on their extensive expertise and interests, the histories embedded in South Asian education systems and in technologies like keyboards, and the overall infrastructure of academic opportunity in contemporary global contexts. All of these factors are critical to understanding the various dynamics at play in international, multilingual research, where linguistic histories can reveal important dimensions about what participation, power, and agency can mean for a specific group or community.

# 6

# LINGUISTIC AND LEGAL ADVOCACY WITH AND FOR INDIGENOUS LANGUAGE INTERPRETERS IN OAXACA

In chapter 5, I described how translation activities were incorporated into a participatory design project in Kathmandu, Nepal.[1] In that context, there were stark differences between my participants' language histories and my own, but there were also several intersections in our language practices. Most of my participants at SAFAR in Kathmandu spoke Nepali and had various experiences and strengths communicating in English, so it was my presence as a white Latina who does not speak Nepali that shifted the linguistic dynamics of that particular translation space (Bloom-Pojar 2018).

Moving from Kathmandu to Oaxaca de Juárez, Mexico, in this chapter I extend the discussion of translation as a participatory design tool even further, outlining a project in which participants who speak over 200 languages and language variants came together with the common goal of supporting Indigenous language rights and representation. Through a discussion of how Indigenous language interpreters and translators negotiated linguistic and cultural differences to design an event that brought attention to the importance of Indigenous language translation and interpretation across the world, I present strategies for negotiating large-scale language differences in participatory design and user experience projects. Through grounded examples of how this transnational, multilingual research team navigated language relations to develop a project focused on Indigenous language support and advocacy, I continue to argue that technical communicators, as techno-scientific rhetoricians, can amplify their impact and potential for social justice (Jones 2016; Jones, Moore, and Walton 2016) by leveraging the role language diversity plays in contemporary technical communication work.

1. Portions of this chapter were previously published in Gonzales 2016 and García et al. 2022.

https://doi.org/10.7330/9781646422760.c006

Perhaps most important, participants in this chapter demonstrate that language cannot and should not be abstracted from the bodies and lands on which it lives and through which it travels and that embracing intersectional, interdependent approaches to multilingual technical communication requires added attunement to the colonial underpinnings of contemporary translation and interpretation work. As you read the stories of the Indigenous language translators and interpreters presented in this chapter, I encourage you to envision what it might look like to practice multilingual technical communication through social justice–driven perspectives that foreground Indigenous knowledges.

## THE INTERNATIONAL YEAR OF INDIGENOUS LANGUAGES

The United Nations Educational, Scientific, and Cultural Organization (UNESCO) declared 2019 the International Year of Indigenous Languages, noting that Indigenous languages are disappearing across the world "at an alarming rate"—particularly due to discrimination in the areas of migration, health, and displacement, among others (UNESCO 2019). That same year and every year, Indigenous communities around the world continue to face violence, theft, and state-sanctioned displacement by colonial governments (Lara 2017). In Mexico, according to the Mexican Political Constitution, Indigenous Mexican communities have a right to express themselves in their own language in every legal transaction, meaning that Indigenous community members in Mexico should be accompanied by interpreters at all legal proceedings (Constitución Política 2017; Lara 2017). However, research conducted by non-government-affiliated organizations such as the Centro Profesional Indígena de Asesoría, Defensa, y Traducción (CEPIADET) (Professional Center for Indigenous Counsel, Defense, and Translation), the organization I introduce in this chapter, shows that approximately 100 percent of the Indigenous non-Spanish-speaking population of Mexico is not provided with an interpreter during their arrest, at their hearings, or during their trials (Kleinert and Stallaert 2015, 238). As this research suggests, there is a vast need for support of Indigenous language representation, sustainability, advocacy, and revitalization in and beyond the Mexican settler states.

In 2018, while I was living and working at an institution located on the Mexico/US border—at a time when, as Zapotec interpreter and activist Odilia Romero explains, migration at the border and the struggle for Indigenous language support in the US continued to develop into "something completely different" than what had been previously experienced (Medina 2019); when, despite the heightening need for Indigenous interpreters in immigration courts at the border, the state of Texas still did not have a single certified court interpreter who worked in Indigenous Mexican or Central American languages (Texas Office of Court Administration 2019); and when Indigenous asylum-seeking children and families were (and still are) detained and jailed at the border in life-threatening conditions at an alarming rate—one of my brilliant students and friends, Nora Rivera, approached me to collaborate with her and several colleagues in different countries to plan a gathering of Indigenous language interpreters and translators in collaboration with CEPIADET. The purpose of this gathering was for Indigenous language interpreters and translators to co-design new protocols and methodologies to support professional development opportunities that can help sustain translation and interpretation services for Indigenous communities in Mexico and across the globe, particularly during a time of such great need, when resources provided by colonial governments for Indigenous language services are in a continuous and steady decline across the world.

## ABOUT CEPIADET

The Centro Profesional Indígena de Asesoría, Defensa, y Traducción is a non-government organization (NGO) in the state of Oaxaca de Juárez, Mexico, that advocates for Indigenous language rights and representation in the Mexican court system. Established in 2005 by a group of Indigenous lawyers who came together in response to their perceived need for greater representation of Indigenous communities in the legal system, CEPIADET (2020) has continued to grow into an Indigenous advocacy organization currently centered on three major goals:

1. Ejercicio de derechos: Coadyuvar en la construcción de instituciones y procedimientos que permitan el pleno ejercicio de los derechos de los pueblos indígenas.
2. Acción y política pública: Evaluar la forma en que las instituciones gubernamentales atienden los problemas individuales y colectivos de las sociedades indígenas y a partir de ellas generar propuestas para que las políticas públicas cuenten con pertinencia cultural y lingüística.
3. Fortalecimiento humano colectivo: Acompañar procesos que permitan a las comunidades indígenas, mirarse y actuar como los únicos responsables de sus condiciones de vida, asegurándose de ejercer plenamente sus derechos.

1. Exercising of rights: To contribute to the establishment of institutions and procedures that allow for the full enactment of Indigenous communities' rights.
2. Public and political action: To assess and develop processes for government institutions to attend to the individual and collective rights of Indigenous communities and through these processes to generate protocols for developing public policies that are linguistically and culturally relevant.
3. Collective strengthening of humanity: To support processes that allow Indigenous communities to acknowledge and act independently as the only responsible entities for their standards of living, ensuring that Indigenous communities can fully exercise their rights.

As demonstrated through the organization's mission and vision quoted above, while CEPIADET works primarily to ensure language access for Indigenous language speakers in and beyond Mexico, it also positions language access as part of a broader multilingual experience that encompasses Indigenous rights, political action, cultural relevance, and collective action.

## THE INTERNATIONAL UNCONFERENCE FOR INDIGENOUS LANGUAGE INTERPRETERS AND TRANSLATORS

The collaborations I introduce in this chapter contribute to CEPIADET's shared goals and were conducted between CEPIADET and various researchers across institutions in the US, Mexico, Canada, and Peru. Together, as part of this collaborative work, our team ended up developing what became known as the International Unconference for Indigenous Interpreters and Translators, which brought together 370 interpreters and translators of Indigenous languages in Oaxaca City on August 8 and 9, 2019 (García et al. 2022). Drawing on the participatory methods put forth by an unconference strategy, which advocates for a community-driven approach to solve problems and discuss issues relevant to attendees' interests and expertise (http://unconference.net/methods/), and grounded in non-hierarchical, decolonial approaches to community building, this two-day event consisted of a day of roundtable discussions focused on themes designated by participants followed by a day of collaborative strategy development. The overall goal of this event was to bring visibility to the importance of Indigenous language translation and interpretation while also starting to develop a sustainable network of interpreters, translators, researchers, and activists working for Indigenous language rights.

The structure of this event echoed what Qwo-Li Driskill describes as "decolonial skillshares," which "refer to indigenous rhetorics, pedagogies, and radical practices that ask us to continue our rhetorical (visual, material, performative, linguistic, etc.) traditions as indigenous people, to transform cultural memories for both indigenous and non-indigenous people, and to create spaces for all of us to learn and teach embodied rhetorical practices as a tactic of decolonization" (58). While the concept of an "unconference" was used as an umbrella term to signify the collaborative efforts of our event, the International Unconference for Indigenous Language Interpreters and Translators did not just follow the low-budget or DIY model of Western-based unconference events. Instead, the notion of an unconference served as a general methodology that signified to various audiences how the conversations of the event would develop based on the goals and objectives of the individuals who showed up to be in community on each day. With a focus on sustainability and on establishing conversations and projects that reflected the goals of Indigenous language speakers and activists first, the unconference was planned in conversation among the planning committee and various participants over the course of a year.

Funding for the event was not provided by a single government or academic institution; instead, members of the planning committee applied for and were granted funding from various entities that sought to specifically support Indigenous language activism. Due to this participatory planning and funding structure, members of the planning committee as well as unconference participants had more freedom to design an event that was not guided by academic formalities. Instead, the unconference was an effort to create space for relationality among various members of different communities who came together to build relationships and establish long-term partnerships geared toward Indigenous language activism and representation in and across various sectors. As Driskill explains in the discussion of "decolonial skillshares," "Colonization and genocide in the Americas and elsewhere depend on the destruction of cultural memory through attacks on indigenous rhetorical practices. To counter these attacks, indigenous people in the United States and Canada [and, as this chapter demonstrates, in Latin America] are in the process of reasserting the importance of indigenous traditions, languages, and knowledges through community-based events" (57). The unconference described in this chapter, then, is just one event among a constellation of efforts the people represented in this project continue fostering to push for Indigenous language rights across the globe.

## CEPIADET'S INTRODUCTION TO THE UNCONFERENCE

Before discussing my own positionality and role in this project, in this section, members of the CEPIADET team—Tomás López Sarabia, Edith Matias Juan, Abigail Castellanos García, Elena García Ortega, and Gaby León Ortiz—provide their overview of and introduction to the unconference event. This reflection was written in Spanish, with portions in Ayuuk/Mixe Media–Santiago Atitlán, Tu'un savi/Mixteco, and Dixhzaa/Zapoteco. I provide an English translation as a footnote.[2]

---

2. This chapter details the development and results of the International Unconference for Indigenous Language Translators and Interpreters, which took place in the city of Oaxaca de Juárez on August 8–9, 2019. The goal of this unconference was to reflect on current issues and established processes for training Indigenous language interpreters and translators and to analyze the quality of these services in various areas of the public sphere, specifically from the perspective of interpreters themselves, as well as through the perspective of public servants, civic employees, and academic organizations. The ultimate goal of this event was to generate strategies and work plans to continue strengthening public policies regarding Indigenous language rights.

   This event was developed through the frame of the International Year of Indigenous Languages, which was designated as such by the United Nations Educational, Scientific, and Cultural Organization (UNESCO) to recognize historically marginalized languages and to raise awareness in the general public about the endangerment of these languages and their role as cultural vehicles, systems of learning, and ways of life—particularly for Indigenous communities, for which languages play a crucial role in public, economic, and cultural participation in any given country.

   The unconference correlates with the annual goals put forth by UNESCO, including that of "increasing access to information in Indigenous languages regarding the role that academic circles and public organizations can play in language preservation, increasing access and support for Indigenous languages, [and] developing sustainable strategies for peace and celebration through Indigenous languages and creative artistic expression" (UNESCO 2019, n.p.). The direct participation of trained Indigenous interpreters is critical to achieving any of these goals.

   The organizing committee of this event was coordinated by the Centro Profesional Indígena de Asesoría, Defensa, y Traducción, Asociación Civil (CEPIADET) (http://cepiadet.org), an NGO made up primarily of bilingual lawyers from various Indigenous communities in the state of Oaxaca—all of whom work to protect the rights of Indigenous language speakers through translation and interpretation services, educational opportunities, and professional development practices—in collaboration with academics at the University of British Columbia in the Creative and Critical Studies Department, the TRACE Innovation Initiative in the University of Florida's Department of English, the Universidad Veracruzana, the University of Texas at El Paso, as well as the Juan de Cordova Research Library. Members of the organizing team developed a partnership grounded in research regarding language rights in Indigenous communities and in previous work CEPIADET has conducted since 2006 in the area of Indigenous interpreter training. This project then also gained input from the Mixteco Indígena Community Organizing Project (MICOP) (http://mixteco.org) in Los Angeles, California, USA, as well as from faculty in the Interpretation and Translation program at the Universidad Peruana de Ciencias Aplicadas (Peruvian University of Applied Sciences).

   The unconference was hosted in the city of Oaxaca, given the fact that this Mexican state has the largest population of Indigenous language speakers in the country

Tyatë nëkyëj yë'ë myëët äjtypy pëntii ojts yäjk nëkäjpx yäjk nëmatyä'ä ëts tii ojts kyopk täny mo tojts tyuyoyën jä'ä käjpxën matyakën "Nëë ijty kyäjpnkëëxm ëts jä'ä ääw ayuuk nyajtëkmujkmëty mëtipe yäjk janyajxpën yäjk käjpxnäxpën (DIITLI), tyuyo'oy Mo wäjkwimëtë käjpn, jaa 8 ëts 9 ämp agostë 2019. Mëët kyëxmpë jä' ä käjpxy mätyä'äk ojts 'ity ëts yäjk

---

and based on the great need to normalize language diversity in the state to allow Indigenous language speakers to be able to use their languages in all capacities, as articulated by Article 9 of the general language law—a law that is still currently not completely applied and upheld.

Those of us who took part in this event believe it's important to go beyond short-term training programs for interpreters and to launch advanced training programs in Indigenous language translation and interpretation, including graduate degrees that can be established, sustained, and assessed long term. This will allow Indigenous communities to have increased access to justice and enjoy the full execution of their human rights while also continuing to position the preservation of language diversity as part of the pluricultural foundation of the Mexican nation and the cultural heritage of humanity writ large.

Through the discussions and analyses across four academic institutions and twelve organizations, this event was based around four main themes:

Developing Strategies for Raising Awareness for Public Officials
Interpreter Training and Professionalization
Translator Training and Professionalization
Managing Indigenous Language Interpretation and Translation Services in the Public Sector

The roundtables at this unconference provided a space to put into perspective various experiences participants have gathered for more than ten years, as organizations, interpreters, translators, and government and academic organizations developed strategies for addressing human rights violations—including language rights violations—inflicted on Indigenous communities. The work of these participants across the years has transformed the landscape for Indigenous communities seeking access to justice.

Through the conversations at each roundtable, the following goals were developed:

Establish a legally recognized network of interpreters and translators for events related to training and employability, especially among Mexico, Peru, and the United States
Share experiences related to interpreter and translator training at both national and international levels
Train interpreters and translators through critical approaches
Develop comprehensive strategies for teaching Indigenous languages
Conduct research on Indigenous interpreter and translator training
Streamline existing data on these issues
Define and streamline payment processes for Indigenous language services in the public sector
Develop advocacy strategies to ensure that institutions get the economic resources needed to pay for Indigenous language translation and interpretation services

We hope the results and proposals generated at the unconference will continue to have a sustainable impact and that we can continue to see advances toward our goals at future gatherings.

mëmay yäjk mëtajëty so tu'un yäjk päättë jä'ä 'iëxpëjkën jä'ä mëku'uk tëjkëty mëtipë jä'ä ääw ayuuk kyapxnäjxtëp jyanyäxtëp ëts näy tu'un so jä'ä jëntsëntëjk tyunk tyäjk tuyo'oytyë; mëëtpë jä'ä 'iääw 'iäyuuk mëtipë kyajpxnäjxtëp amäxän—ayuuk—amaxän, jëntsën tëjkëty, yäjk ëxpëjpë tëjkëty ëts jä'ä ja'yety mëtipë tukmuk tuuntepën ëts nëtëkmuk tyäjk tuyo'oytyëty tyatuunkpe tu'unën pëjkën.

Loo kitz re rxha'g za'a ira' xhte xa guuk ne xa gue ya'n irate ni guuk loo "Gal rxa'g za'a mniit ni rtuid de dixh xhte de gxh baañ" ni guuk Lua' xihini gul xuun ne ga' xhii beu agosto xte 2019. Ni buiñ guuk tziñki naa te guk xgaab xa zezaa gal rzuid xte de mniit ni rtuid de dixh ne la gizla' rak tziñki lat rki'ñni; bxag za' xgaab xte de mnit ni rtuid dixh, xte mniit ni ruiñ tziñ par ubiern, xte de mniit ni rxag za'ni te gaaknedeb de mniit, xgaab xhte mniit ni rzuid; te gukxe' xa gaak guiñdeb tziñ iraadeb ne guiñdeb tio'ru irate ni a ka yuiñ ubiern nare.

Nu´u tutu yo vasi ndi´i ña ni satiun na saya´a tu´un saá si tu´un savi, ña ku ki´i una si iin agosto kui uvi mili siuan kumi ni sio ña ku induva. Ni ndandukuña na ndatu´una sia ndatyi saanana si ndatyi satiun na saya´a tu´un saá si tu´un savi, tava na ku ndetaá tu´un na ndika tisi gobierno, na sana´a, si na satiun ndatyi ku tyinte´ena sia ña.

Este capítulo recopila el desarrollo y los resultados de la Desconferencia Internacional de Intérpretes y Traductores de Lenguas Indígenas (DIITLI), realizada en la ciudad de Oaxaca de Juárez, los días 8 y 9 de agosto de 2019, con el objetivo de reflexionar en torno al estado actual de los procesos de formación de intérpretes de lenguas indígenas y a la calidad de los servicios en el ámbito público desde las perspectivas de los intérpretes, servidores públicos, organizaciones de la sociedad civil y académicos, a fin de generar estrategias de trabajo articuladas que fortalezcan las políticas públicas en la materia.

El evento se desarrolló en el marco del Año Internacional de Lenguas Indígenas, decretado así por la Organización de las Naciones Unidas para la Educación, la Ciencia y la Cultura (UNESCO) y dedicado a las lenguas históricamente marginadas a fin de sensibilizar a la opinión pública sobre los riesgos a los que se enfrentan estas lenguas y su valor como vehículos de la cultura, los sistemas de conocimiento y los modos de vida, reconociendo que las lenguas indígenas desempeñan un papel crucial para que las comunidades de hablantes asuman su destino y participen en la vida económica, cultural, y política de sus países.

En dicho sentido, la DIITLI empata con las líneas temáticas propuestas por la UNESCO puesto que abarcó temas que van desde el acceso a la información en lenguas indígenas al papel y la contribución de los círculos académicos y las organizaciones públicas en la preservación, el acceso y el apoyo a estos idiomas, los idiomas indígenas, el desarrollo sostenible y la paz, y la celebración de las lenguas indígenas mediante la expresión artística y la creatividad (UNESCO 2019, n.p.). Y para lograr dichos planteamientos se requiere la participación de intérpretes profesionales.

El comité organizador de este evento, coordinado por el Centro Profesional Indígena de Asesoría, Defensa y Traducción, Asociación Civil (CEPIADET)—una organización no gubernamental (ONG) compuesta principalmente por abogados bilingües de varias comunidades indígenas del estado de Oaxaca que trabajan para proteger los derechos lingüísticos de los hablantes de lenguas indígenas a través de los servicios de interpretación y traducción, oportunidades educativas, y prestación de servicios profesionales—e integrado por docentes de la Universidad de Columbia Británica de la Facultad de Estudios Críticos y Creativos; The TRACE Innovation Initiative del Departamento de Inglés de la Universidad de la Florida; La Universidad Veracruzana, La Universidad de Texas en El Paso, y la Biblioteca de Investigación Juan de Córdova, generaron una alianza que tiene como antecedente los trabajos sobre difusión de derechos lingüísticos de los pueblos indígenas y los procesos de formación de intérpretes que CEPIADET ha desarrollado desde el año 2006. A este esfuerzo se sumaron integrantes del Proyecto Comunitario Mixteco Indígena con sede en Los Ángeles California, Estados Unidos, y la Facultad de Traducción e Interpretación de la Universidad Peruana de Ciencias Aplicadas.

El evento tuvo lugar en la ciudad de Oaxaca dado que es la entidad en la que se concentra la mayor diversidad lingüística de México y la necesidad de generar y materializar un proceso de normalización lingüística que permita a los hablantes de lenguas indígenas hacer uso de éstas en los términos que enuncia el Art. 9 de la ley general de derechos lingüísticos y que hasta este momento no son realizables íntegramente.

Quienes formamos parte de este evento, consideramos que es necesario trascender los procesos de formación de corto plazo e impulsar otros programas de formación de nivel superior que incluyan postgrados permanentes y susceptibles de evaluación a largo plazo. Lo anterior permitirá garantizar un mayor acceso a la justicia, el goce integral de los derechos humanos de los Pueblos Indígenas y la preservación de la diversidad lingüística como elemento constitutivo de la pluriculturalidad de la nación mexicana y como patrimonio cultural inmaterial de la humanidad.

Tyatë tuunk tyatë kajpx mätyä'äk jä'ä 51 ja'ay, 4 jä'ä ja'yety mëtipë tuuntepën tekmuk ëts 12 winketypya tuuntakn mëtipë ojts yäjk päättë, te tu'un ojts nëkäjpxtë nëmatyä'ä kyë 4 jä'ä ääw ayuuk:

- Yajk tëk äwänëty, yajk tëk mëmatyäkëty jä'ä wëntsëntëjkëty ko jä'ä ääw ayuuk mëtipë yajk käjpxp kopk pëkyëj yë'ë.
- Jyak ëxpëjk ätë'ëtsëty ëts tjäknëjawyë'atë'ëtsëty mëtipë jä'ä ääw ayuuk kyajpxnäjxtëp.
- Yäjk amtow yajk pëjktsowëty mëtipë jä'ä ääw ayuuuk jyay nyäjxtëp kyajpxnäjxtëp mo kyaj jä'ä jä`äyëty ayuuk tkäjpxtë.

Tziñ, gal xgaab ne gal rgue dixh ni buiñ de 51 mniit ni xa'ag lo tap organización, tz'ptiop de instituciones guukni loo de tap dixhre:

- Tziñ te gaak xu giik de mniit ni ruiñ tziñ loo ubiern, de derexh ni rap de mniit inni' xh dixh'ni.
- Gaak gal rzuid te gaak utuid zak' mniit de dixh.
- Gaak gal rzuid te gaak utuid ne ukua zak mniit de dixh.

Gaak tziñ te xu' ni guiñ gal rtuid ne gal rkua' dixh lo irate de liiz ubiern.

Kumi sia´a ña ni ndatu´un uvi siko in na´a, kumi ve´e nu´u satiun na´a, si usi uvi ve´e na gobierno:

- Ndatyi ku sa´ana, na kua´a si ndasamana nuu ta´ana, tutu tava na kundani na ndika tiun gobierno
- ndatyi ku sana´ana sanana nu ú na saya´a tu´un sa´a si tu´un savi
- ndatyi ku sana´ana sanana nu´u na ta´a tava na saya´a tu´un sa´a si tu´un savi
- ndatyi sa´ana sikana na satiunna na saya´a tu´un sa´a si tu´un savi

El trabajo de análisis y debate que llevaron a cabo las personas, asociaciones, y instituciones que participaron en este evento, versó sobre cuatro temas:

- Mecanismos de sensibilización a servidores públicos e incidencia sobre los derechos lingüísticos.
- Formación y profesionalización de intérpretes.
- Formación y profesionalización de traductores.
- Gestión de los servicios de interpretación y traducción de lenguas indígenas en los servicios públicos.

En las mesas de trabajo se pusieron en perspectiva las experiencias que a lo largo de más de diez años las organizaciones, las y los intérpretes, las y los traductores, así como algunas instituciones académicas y gubernamentales han generado para atender la violación a los derechos humanos de los Pueblos Indígenas, particularmente de los derechos lingüísticos, que han permitido incidir para cambiar los escenarios de acceso a la justicia para los Pueblos Indígenas.

De las mesas de análisis se formularon las siguientes propuestas:

- Crear una red con personalidad jurídica de intérpretes y traductores para las incidencias en temas de formación y empleabilidad, especialmente entre México, Perú y los Estados Unidos
- Compartir experiencias de formación a nivel nacional e internacional
- Formar traductores e intérpretes desde una perspectiva crítica
- Generar estrategias integrales para la enseñanza de las lenguas
- Realizar investigaciones específicas sobre el tema
- Sistematizar datos
- Definir criterios de pago para los servicios de traducción e interpretación en los servicios públicos
- Generar estrategias de incidencia para que las instituciones cuenten con recursos económicos para el pago de los servicios de interpretación y traducción

Yajk äxijtp ëts tyatë tu'unk nëky mo ojts jyënkëtakyën tyatë ääw ayuuk tyunëty ëts ko winkpë yajk nëpëmëty yajk nëpëjktakëty jä'ä käjpxy mätyä'äky yajk ixëty yajk tunëty tyatë mëëtjawi'inpë nëkyëj.

Ka beezne irate b'iñ ni bluikne gak zakni ne nare ni xizaa ixa'gne ugui'n ni irikaa lo xhtziñne.

Ndatiundiu ña ndi´i ña ni ni sa´andiu na tyinteña tava na k uta na ndetandiu tyi nu´u, na ku kunde´endiu tiun.

Esperamos que los resultados y la articulación generada nos rinda frutos y que las siguientes reuniones podamos vislumbrar resultados sobre las acciones propuestas en este primer encuentro.

## RESEARCHER ROLE AND POSITIONALITY

As a technical communicator, researcher, translator, and interpreter who takes a rhetorical approach to studying multilingualism, I was invited to collaborate with CEPIADET on the development of this unconference, contributing to the overall strategy of the event while also helping strengthen the sustainability of our collaboration after the event—specifically by co-designing technical documentation that can be used to foster future projects and collaborations among event attendees. While CEPIADET's main goal in organizing this event was to bring together diverse perspectives on Indigenous language translation and interpretation services across the world, as Tomás, a founding member of CEPIADET, explained during one of our planning sessions, "Este evento no es solo un rosario de quejas—si no es el principio de una gran intervención" ("This event is not a 'rosary of complaints' but is instead the start of a great intervention"). In referencing the metaphor of "a rosary of complaints," Tomás alluded to the fact that Indigenous languages and Indigenous communities across the world continue to face ongoing discrimination; while the naming and processing of this discrimination are important, as Tomás also clarified, the purpose of CEPIADET's gathering was also to develop practical tools and strategies for working through this discrimination to find solutions and develop ways to support Indigenous communities through purposeful coalitions (Jones, Moore, and Walton 2016; Jones 2020) and technical interventions.

As a South American immigrant originally from Santa Cruz, Bolivia, I experienced the impacts of this collaboration on both a personal and a professional level. During the time of my ongoing collaboration with CEPIADET, Indigenous communities in my home country suffered severe state-sanctioned violence during the 2019–2020 coup d'état in Bolivia (Swain 2019) that echoes the violence and persecution Indigenous communities suffer in Bolivia and throughout the world. Thus, becoming involved in this collaboration was important to me not only as a researcher and technical communicator but also as a Bolivian seeking to learn how to better support Indigenous liberation movements. To ignore these events as well as my own positionality as a white Bolivian who does not speak an Indigenous language and who theorizes the connections among rhetoric, language, and technology would prevent me from engaging fully in the intersectional, interdependent methodology I propose in this book. As I outline in the next section, tracing and practicing multilingual technical communication—for the purposes of this collaboration as well as in all technical communication work—requires an attunement to power, positionality, and privilege (Jones, Moore, and Walton 2016) as well as a keen awareness of how language relations shape multilingual, transnational research, particularly with Indigenous communities (Itchuaqiyaq 2021).

## PRACTICING DECOLONIALITY AND PARTICIPATORY DESIGN WITH CEPIADET

As a contributor to CEPIADET's event planning team, my role in this collaboration was to co-develop protocols and strategies for fostering conversation and participation among the 370 attendees who came together for this gathering, with the goal of developing tools and strategies for supporting the work of Indigenous language interpreters and translators in Mexico and across the world. Attendees at this event included Indigenous language interpreters and translators from

Mexico, Peru, Canada, and the United States, as well as academic researchers, students, and community members. In addition to helping with the event strategy and project management, I was tasked with co-developing technical documentation, including written materials to be shared both during and after the event to illustrate the work plan developed during the gathering as well as next steps for this collaboration.

As Bruce Maylath and Kirk St.Amant (2019) explain, "Technical communicators are the ones who develop the texts explaining how a process works or how a technology operates" (3). Therefore, technical communicators are also "well equipped to work with translation and localization professionals," particularly in contemporary global contexts (3). At the same time, it's important to also embrace ethical, justice-driven frameworks for engaging in collaborations between translators and technical communicators. In this project, to successfully complete the tasks CEPIADET had asked me to do, we embraced a decolonial methodology that could help position CEPIADET as well as event attendees as experts in their own communities who possess talents and skills that I, as a person who does not speak an Indigenous language, do not possess. For this reason, in framing my goals and tracing my process through this collaboration, my intersectional, interdependent orientation to multilingual technical communication work requires that I thread together decolonial methodologies and participatory design.

Definitions and applications of decoloniality and decolonial methodologies vary widely across fields. In technical communication specifically, Indigenous scholars bring attention to decolonial methodologies that center "flexibility, reflexivity, humility, and respect" as grounding principles for technical communication research and practice (Agboka 2014, 316). As Godwin Y. Agboka (2014) clarifies, "Current approaches to both researching and designing technologies are [often] motivated by modernist ideologies . . . whose history is tied to the colonial project" (298). Rather than focus on models of expediency and efficiency that are at the roots of technical communication's history and its ties to the industrial revolution (Jones 2016), as Angela M. Haas (2012) reminds technical communication researchers, doing decolonial work means we must "support the coexistence of cultures, languages, literacies, memories, histories, places, and spaces—and encourage respectful and reciprocal dialogue between and across them" (297). In this way, a decolonial orientation to multilingual technical communication inherently rejects the notion that languages are discreet, stable, and neutral and that communication can happen across languages without consequence or impact on the lands, spaces, and people who are implicated in this communicative practice (Itchuaqiyaq 2021). An intersectional, interdependent methodology aligns with decolonial principles, particularly by recognizing the connections among all stakeholders in a project as well as their surrounding communities, environments, and lands.

As Indigenous language researchers and activists have long argued, decolonial approaches to language diversity must "visibilizar cómo el elemento cultural 'lengua' es un concepto polisémico que en este campo refiere más a sistemas de representación social y de organización del pensamiento que un sistema de sonidos y/o palabras" ("make visible that the cultural element known as 'language' is a polysemic concept that [in the field of Indigenous language studies] refers more to a system of social representation and organization of thought than to a system of only sounds and/or words") (Hernández 2019, 19). As Indigenous language researcher Lorena Córdova Hernández (2019) explains, Indigenous language research and Indigenous language revitalization efforts work toward "la movilización social [que] sea para conseguir la (re)conquista de espacios (fisicos y virtuales) que permitan el restablecimiento del uso y transmission de un sistema comunicativo y de conocimiento" ("social mobility as the reclamation of spaces [both physical and virtual] that will allow for the reestablishment and use of a communicative system and way of knowing") (19). In other words, a decolonial orientation to multilingual technical communication research and to language work more broadly should emphasize language in relation to—rather than try to abstract language from—the people, cultures, relations, and lands on and through which linguistic practice is enacted. In this way, a decolonial orientation to multilingual technical communication research can function through intersectional, interdependent lenses that push for fluid rather than binary definitions of language and culture.

To work with CEPIADET through decolonial methodologies that embrace decoloniality as more than a metaphor (Tuck and Yang 2012), our research team and unconference planning committee decided to thread decolonial orientations with participatory design, specifically to provide tangible methods and practices for non-Indigenous members of our research group to center the perspectives of the Indigenous communities and Indigenous language speakers with whom we collaborated on this project. As Emma Rose and Alison Cardinal (2018) explain, while participatory design can be "essential in TPC work" as a way to "do social justice work . . . collectively and in solidarity with people and groups that are currently marginalized" (20), the notion of participatory practices and design is at the core of Indigenous epistemologies that have long advocated for collective, community-driven decision-making and governance (Simpson 2017; Smith 1999). As Mónica Morales-Good (2022), a member of our unconference planning committee, explains in her discussion of the unconference event we co-hosted with CEPIADET, "Increasingly, more emphasis is placed on the value of decolonization in relation to issues of language and translation. Indigenous communities actively participate in this process of decolonization, in order to ensure that Indigenous spaces, perspectives, and knowledges are centralized. After all, decolonization cannot be achieved without the active participation of Indigenous communities—those who embrace a responsibility to recognizing the voices of our pasts and the knowledge of our elders in order to work with current governments to preserve and reclaim Indigenous territories, languages, as well as cultural and political customs that have survived through generations" (n.p.). For the purposes of this project, our team combined participatory design as it is practiced in technical communication and technology design with Indigenous epistemologies and methodologies as a way to thread (rather than superimpose) orientations and applications of participatory collaborative practice across our team. The goal here was not to "bring participatory design" to the Indigenous language speakers and activists involved in our collaboration but rather to find a point of intersection and interdependency among our research team on the idea that participation (with an attunement to power differentials and established social structures) from all parts of our team was essential to the success and sustainability of our work together.

Although my collaboration with CEPIADET is ongoing and has transformed throughout the years, in this chapter I report on data that were gathered during the unconference event in Oaxaca City in August 2019, where attendees participated in four roundtable discussions as well as an opening and closing panel over the course of two days. In addition, I also draw on field notes I took over the course of twelve months of collaborating with CEPIADET, both in planning the unconference event and in writing the documentation following the unconference. In these field notes, I included quotes from virtual and in-person meetings as well as emerging thematic patterns on how members of my team and I negotiated translation during our meeting and planning. In keeping detailed field notes of our meetings, my goal was to document how language diversity shaped this collaboration and the decisions the team made when planning an unconference event and when drafting documentation from the unconference for future use.

To provide in-depth analyses of these data in a contextualized way, in the sections that follow, I elaborate on four major themes that emerged from the corroborated analysis of both the audio recordings of the unconference event and my field notes: (1) validating the presence and value of language difference, (2) connecting language and land, (3) challenges and affordances of multilingual technical documentation in Indigenous languages, and (4) co-designing technologies to support Indigenous sovereignty. In describing these themes in the sections that follow, I offer examples and focus primarily on implications for technical communicators interested in doing multilingual technical communication work in global contexts.

## VALIDATING THE PRESENCE AND VALUE OF LANGUAGE DIFFERENCE

Many discussions of translation and interpretation are focused on the importance of ensuring that everyone involved in a particular communicative event has access to all the information being shared. For example, in professional interpretation contexts, there are two types of common interpretation practices: simultaneous interpretation and consecutive interpretation. Simultaneous interpretation takes place when an interpreter converts what is being said in one language into another language in real time, as a person or presenter is speaking. Simultaneous interpretation is typically used during large events, where individuals who want to access the information presented on a stage will listen to an interpreter through a headset. There is no pause between the time when the speaker is talking and when the interpreter speaks. This tends to save time and allows attendees to hear information in their preferred language in real time. Consecutive interpretation, in contrast, requires that a speaker pause to allow time for the interpreter to translate information. Consecutive interpretation is often practiced in small group or one-to-one interactions, such as a doctor's visit, where an interpreter facilitates communication between a patient and a healthcare provider who speak different languages.

In academic research, both simultaneous and consecutive interpretation are common practices, allowing researchers and participants to communicate across languages in research activities such as interviews, focus groups, and presentations. While both types of interpretation could have worked in some capacity during CEPIADET's unconference event, the vast number of Indigenous languages represented, as well as the lack of funding to pay designated interpreters to interpret, led to both challenges and possibilities regarding interpretation. While some of these challenges and possibilities may be unique to this particular event, the strategies unconference attendees used to mitigate communicative discrepancies may have potential implications for other multilingual technical communication projects that encompass rich linguistic diversity and limited resources.

Participants at CEPIADET's unconference spoke more than 200 languages and language variants collectively. All attendees identified as having at least some proficiency in Spanish, so a majority of the conversations were Spanish-dominant. At the same time, however, at a conference emphasizing language diversity, participants exhibited several strategies that both incorporated and resisted Spanish as the assumed "neutral" language of conversation, and they found ways to incorporate Indigenous languages into the discussion without translation or interpretation into Spanish. Like participants in Kathmandu who used Nepali and Rai languages in their communication during our collaborative work, participants at CEPIADET's unconference event leveraged their linguistic histories to establish and make space in and beyond this event.

For example, the 370 participants at the unconference event were separated into four roundtables. Each roundtable had a specific theme that was decided upon by the participants themselves. As members of CEPIADET mentioned in their reflective introduction, these themes included professional development strategies for Indigenous language interpreters, professional development strategies for Indigenous language translators, collaborating with public officials on effective Indigenous language translation and interpretation, and identifying and creating job opportunities for Indigenous language interpreters and translators in the public sphere. While each roundtable had a designated moderator and set of roundtable participants and the conversations at each table differed greatly, each roundtable discussion began organically in the same way: with each participant introducing themselves by stating their name, what community/communities they are from, and the languages they speak.

While these introductions may be nothing unusual for this type of gathering, participants who are Indigenous language speakers introduced themselves in various Indigenous languages, even as they understood that other members of the roundtable may not have been able to understand what was communicated during the introduction. For the Indigenous language speakers in the room, the purpose of using their Indigenous language(s) to introduce themselves was to, as one participant explained in the discussion, "make our languages present in this room and at this event." While participants knew how to introduce themselves in Spanish, they introduced themselves in their heritage Indigenous languages to make space for those languages in the discussion by requiring that all attendees spend time and energy listening.

This type of introduction, connecting language histories and Indigenous languages to specific communities, is a common practice for Indigenous language speakers, for whom language and community are always interconnected (Driskill 2015; Loebick and Torrez 2016). As Driskill (2015) explains, "Decolonization commits itself in solidarity by acknowledging and engaging with both the commonalities and the differences in experiences of oppression and struggles for change" (75–76). Making space for everyone's language during the introduction, then, meant making space for both our commonalities and our differences. Furthermore, including everyone's language in the introductory portion of the unconference contributed to the decolonial skillshare practiced at the event, where participants taught each other pieces of their languages and therefore of themselves through Indigenous language sharing. Driskill continues, "Teaching indigenous languages—even if minimally—is a deeply radical act" (68). Because "languages are not something human beings *have* but what human beings are," participants' languages are a central part of their introductions, histories, and work and thus should have space in collaborative events, even when not all words are accessible to all people in the same way (Tlostanova and Mignolo 2012, 61, original emphasis).

As the unconference continued beyond introductions, participants continued sharing dialogue in their Indigenous languages as they contributed to roundtable discussions. For example, at one of the roundtables, Shara, a certified Quechua-Spanish interpreter from Peru, followed her brief introduction by speaking in Quechua for several minutes. As Shara spoke, she paused at several points to allow participants to truly hear her and her language before she transitioned to speaking Spanish.

As the only Quechua speaker at the roundtable, Shara knew other participants would not understand the words she was saying, but she stated that she wanted to ensure that her language, like her body, was present in the room. Following her discussion in Quechua, Shara continued in Spanish, stating, "Nadie ama lo que no conoce, Nadie da de lo que no tiene, Nadie enseña de lo que no sabe. Hablar nuestras lenguas indígenas, es ejercer el principio de nuestra identidad, con dignidad, orgullo, y soberanía cultural" ("Nobody loves what they don't understand, nobody gives what they don't have, and nobody teaches

*Figure 6.1. Shara Huaman Julluni introduces herself.*

what they don't know. Speaking our Indigenous languages means exercising the principles of our cultural identity with dignity, pride, and cultural sovereignty").

While Shara moved on from this brief introduction relatively quickly to focus on describing a radio program she hosts for Quechua speakers in Peru, her brief Quechua introduction as well as her translated statement "Nobody loves what they don't understand . . ." signaled Shara's insistence on making space for Quechua in every space she enters—regardless of whether others in her presence understand

Quechua. This same process was echoed by several participants at the roundtables, who chose to speak in their heritage languages without always providing translations. In these cases, the focus of the discussion was not on ensuring that everyone understood every word that was said in the conversation but rather that all participants in the room felt as though their languages were present and represented at the event. For the participants at the unconference, simply stating that they speak a specific Indigenous language was not enough representation. Instead, participants wanted their language to be heard in our shared space—to take up room in the discussion without having to be translated or explained.

In my analysis of the roundtable conversations, it was interesting to note not only how and when participants spoke in Indigenous languages but also how other participants listened and engaged with Indigenous languages, even if they did not understand the specific words that were spoken. For example, although Shara was the only Quechua speaker at her roundtable, when she spoke in Quechua, other roundtable participants listened intently by nodding, humming small sounds of approval, and thanking Shara for sharing her knowledge in her language. While these moments may be classified as "translation moments" (Gonzales 2018) or translation events during the discussion because they required the negotiation of meaning across languages, this negotiation of meaning did not take place through typical interpretation or translation protocols. Instead, translation in this case took place through the mutual understanding that all Indigenous languages deserve space in the conversation and that not everyone in the room will have access to every word being spoken. The role of language difference in these cases was to establish space and relationality, as participants acknowledged each other's languages without needing to have everything translated into a "neutral" or common language or the default colonial language (i.e., Spanish) that was sanctioned at this particular meeting. In this way, multilingual experiences were designed and sustained in this space not through translation alone but rather through the mutual respect and responsibility of honoring languages and making space to recognize, acknowledge, and listen for language difference.

## CONNECTING LANGUAGE AND LAND

Shara and other participants' emphasis on "creating space" for their Indigenous languages at the unconference extended to a second theme that emerged through my collaboration with CEPIADET: the connections between language and land. In "Decolonization Is Not a Metaphor," Eve Tuck and K. Wayne Yang (2012) emphasize what Indigenous communities have been arguing for centuries—namely, that decolonization should be practiced in direct relation to land repatriation. According to Tuck and Yang (2012), many academic researchers use the term *decolonization* "as a metaphor for other things we want to do to improve our societies and schools" without making explicit connections to land and the displacement of Indigenous peoples (1).

During the unconference event, participants reiterated the need to emphasize land in discussions of Indigenous language praxis, including but not limited to Indigenous languages and their translation and interpretation. Through this discussion, participants highlighted the importance not only of land repatriation but also of recognizing how land, like language, consistently shifts and changes. Because land, like languages and people, is not static, decolonial activists and scholars should recognize not only the land that was stolen from Indigenous communities but also how communities have been and continue to be displaced while also migrating across lands voluntarily or involuntarily through enslavement, deportation, and more (Garba and Sorentino 2020).

Through an attunement to the connections between language and land, participants at the unconference affirmed Tuck and Yang's (2012) emphasis on land as a central element of decolonization while also echoing other scholars, such as Tapji Garba and Sara-Maria Sorentino (2020), who emphasize the importance of recognizing not only the stealing of material land but also the stealing and displacement of people across these lands. While Garba and Sorentino (2012) are talking about enslavement and anti-Blackness specifically, unconference participants discussed the connections between land and language by emphasizing the fact that people, languages, and lands are intertwined and should all be considered in decolonial activism. In other words, for participants at the unconference, language activism requires attunement to the stories and histories that shape and are shaped by lands, languages, and people across time.

During the opening session of the unconference, the director of CEPIADET, Tomás, welcomed all attendees and explained that the goal of our gathering was much broader than what could be achieved in two days. The fight for language justice should always be tied to decolonization, Tómas explained, because "our languages need a place to land. All these conversations about language revitalization and growth are great, but our languages can't be revitalized only to float on thin air. Without the reclamation of land, the full reclamation of our language is impossible. Our language and our land are one."

Tomás's reference to languages "float[ing] on thin air" echoes Tuck and Yang (2012) as well as many other Indigenous studies researchers and activists, who explain, among other things, that decolonization efforts should have grounding in the physical reclamation of land and territory and that Indigenous language practices are inherently connected to the lands and bodies in which these languages reside (Ríos 2015). Without an attunement to land, conversations about translation and interpretation will also ignore the bodies that make translation and interpretation possible, the journeys interpreters and translators have to undertake to complete their work, and the displacement of people that often perpetuates the need for translation and interpretation in the first place.

During a roundtable discussion, one participant, Misael, originally from the municipality of Mazatlán, Villa de Flores, Oaxaca, and an interpreter of Mazateco to Spanish and vice versa, described his experience interpreting for speakers of Mazateco in the Mexican courts by discussing the great distance physically, metaphorically, and emotionally he frequently travels to complete his work: "Typically, in our Indigenous communities, there is a local authority. Members of this authority speak your language, understand your customs and your traditions, because they are from the same community. But when an Indigenous brother comes to a court hearing facilitated by the state government, instead of feeling confident, he feels scared, terrified, and no matter how much he may want to express what he agrees with or doesn't agree with, he won't do it. Why? Because of the space he's in and the people he's around."

Misael went on to explain that in order to interpret for his Indigenous community, he has to travel 386 kilometers from his home municipality to Mexico City, which often takes ten–twelve hours by bus and costs Misael several hundred uncompensated Mexican pesos. Upon arriving at the hearing to conduct his interpretation work, Misael meets the defendant, typically a member of his own municipality or community. As Misael explains, one of the biggest challenges in this interpretation work is that although he may be at the court only, as formal interpretation protocols typically claim, to "serve as a conduit of information from one language to another," Misael and the defendant will typically be the only Brown Indigenous people in the courtroom filled with white, Spanish-speaking Mexicans. Misael continued: "They [the courts] usually call us as interpreters to come in at the last minute. We show up to the hearing, and we don't even know the defendant, and there is no time to even meet the person you are there to help. So instead of establishing trust, the defendant also doesn't know why the interpreter is there at the hearing, and he'll often ask, 'Okay, you're here as an interpreter. Does that mean you're here to defend me? Or why are you here?' "

The fact that Mexican state policy sanctions and presumably guarantees the presence of an interpreter for Indigenous language speakers and the fact that both Misael and his defendant are present in the courtroom and speak Mazateco does not, as Misael clarifies, ensure a truly bilingual or multilingual space where all parties can engage in communication. In other words, welcoming or even requiring the presence of bilingual and multilingual speakers in a colonial institution and/or a predominantly white/Western project does not promote or support language diversity. In fact, as became evident through Misael's experiences, the inclusion of an Indigenous language in a colonial context may perpetuate violence, since, as an Indigenous language interpreter, the responsibility of language accessibility falls on Misael's shoulders while Misael himself still has to operate under the Western, colonial infrastructure that dictates that interpreters must remain "neutral conduits of information" and must only speak in Mazateco the words that were initially stated in Spanish by a member of the Mexican state system. Without acknowledging Misael's uncompensated journey to Mexico City or recognizing the displacement Indigenous communities face as they have to travel hours to enter the main city in which the courts are located, discussions of and efforts toward language access remain limited and superficial. The physical travel Misael endures to interpret for his community is one aspect of the violence and oppression he and his community face in order to be recognized and acknowledged in colonial systems within colonized lands.

During a roundtable conversation, two Indigenous language interpreters from Campeche, Mexico, and Quintana Roo, Mexico, discussed various definitions of interpretation. During this discussion, one participant, Mayusa, explained that the presence of interpretation services for an Indigenous language speaker does not guarantee that the speaker's rights are protected. As Mayusa explained:

> En los litigious que hacemos como intérpretes, siempre estamos incorporando la diferencia cultural, tratando de mostrar la diferencia cultural y cuestiones de interculturalidad. Transitamos entre lo que se llama el litigio estratégico que es poner visiblemente los derechos humanos de los pueblos indígenas y ahora estamos tratando de transitar también a lo que se llama un litigio estructural donde mostremos no solamente la falta de un debido proceso como tal, sino todas las violaciones estructurales que están de fondo y que la cedan la integridad de la población indígena, incluyendo nuestros territorios.

> In all the proceedings we conduct as interpreters, we are dealing with cultural differences, and we try to demonstrate these cultural differences and intercultural issues in our work. As interpreters, we advocate for what we call strategic litigation, which means making visible the human rights of Indigenous communities. Right now, we are also trying to develop what we call structural litigation, where we can show not only that there is a lack of due process when it comes to the presence of an interpreter but that there are also various structural human rights violations that are at the core of these processes and that influence the integration of Indigenous communities into [colonial] systems, including violations related to our territories.

In this discussion, Mayusa distinguishes between the presence of an interpreter and the availability of translation and the human rights and adequate representation of Indigenous communities. As she explains, having content translated into or interpreted in Indigenous languages does not mean this content will be culturally appropriate and aligned with the values and orientations of Indigenous communities—communities that have their own methods of governance and continue to be structurally oppressed through the colonization of their lands and territories.

While many technical communication scholars have pointed to the differences between the literal translation of content from one language to another and the localization of information for specific communities, Mayusa, Misael, and other unconference participants extend these distinctions between translation/interpretation and localization by pointing to the specific connections between language and land and the power relationships that continue to be fostered through the structural oppression of Indigenous peoples in Mexico. If translation and interpretation are sanctioned and provided by colonial infrastructures on colonized land, as is the case with the Mexican government in these examples, the presence of language diversity may only benefit these colonial structures while simultaneously and even perhaps unintentionally further contributing to the structural oppression and erasure that has historically deemed some languages inferior to others. Traditional translation and interpretation practices, in addition to the rules and regulations about translation and interpretation established by colonial governments, do not always align with Indigenous language practices and do not always honor the embodied realities and experiences of Indigenous language speakers. This includes, for example, erasing Misael's long journey to complete his work, separating translation and interpretation from cultural and human rights issues, and ignoring the way colonization shapes power relationships in legal proceedings. Thus, as technical communication researchers continue to work with

Indigenous communities and with multilingual communities of color more broadly, it's important to recognize that "welcoming" multilingualism into our research does not mean we are not operating under oppressive colonial ideologies. Attuning to multilingual experiences rather than simply providing or ensuring translation services can help technical communicators recognize the intersectional, interdependent nature of language and its connections to communities.

## THE CHALLENGES AND AFFORDANCES OF DESIGNING MULTILINGUAL TECHNICAL DOCUMENTATION

While most of the translation moments and the broader conversations about language difference at the unconference were focused on verbal interpretation, there were also important conversations about written translations of Indigenous languages that can offer potential implications for multilingual technical communication. Historically, as Linda Tuhiwai Smith (1999) explains, the act of alphabetic writing has been "viewed as the mark of a superior civilization[,] and other societies have been judged, by this view, to be incapable of thinking critically and objectively, or having distance from ideas and emotions" (797). For many speakers of Indigenous languages, alphabetic writing has been imposed as a barrier to language access and sustainability, as colonial governments continue to use written records to erase the presence of Indigenous peoples and Indigenous languages in contemporary contexts.

A major goal of the unconference planning committee was to extend conversations that took place at this inaugural event into ongoing, sustainable initiatives and collaborations among Indigenous language interpreters and translators across Latin America, the US, and Canada. As a first step toward these longer-term initiatives, the planning committee put together a set of unconference proceedings—a written document that would represent the conversations that took place at each roundtable as well as during the closing ceremony of the unconference. One of the biggest challenges we faced in putting these proceedings together was deciding how to incorporate various languages into a single proceedings document: How many languages should we include in the document? Should all content be translated into all the languages? If so, who will do these translations, and how will they be compensated? Who is the target audience for this document, and how will they interact with the document itself?

In discussing these questions with participants at the unconference, one participant, Gloribel, an interpreter from Michoacan, Mexico, explained that Indigenous communities frequently face these tensions regarding written translation and verbal interpretation. As she explained: "Verbal interpretation [in Indigenous languages] is practiced more frequently than written translation, particularly in the areas of health and legal services. However, in this field, we also need to recognize the importance of the written word. A question that typically comes up in these instances is, 'well, who is going to read our languages in written form?' However, writing can actually be part of the revitalization process for Indigenous languages and cultures because if we don't write our words, we don't become part of the written record that colonial agencies value."

In this discussion, Gloribel points to a common assumption regarding the fact that many Indigenous languages are not written in alphabetic form, as they may draw on oral traditions and other forms of meaning making. However, as Gloribel clarified, writing and documenting Indigenous languages is critical to their survival, particularly since written documentation, or the lack of this written documentation, is often used by colonial forces to oppress (through erasure) Indigenous language speakers worldwide.

Through ongoing discussions with both the planning committee and unconference attendees, we decided that it was important to include Indigenous languages in the documentation of the unconference (i.e., the written proceedings). At the same time, we also recognized that incorporating all content in all the Indigenous languages present at the unconference would take much more time and resources than our planning team had at hand, especially if we were to remain as ethical as possible by compensating translators for their time. Thus, based on feedback we received during the unconference closing session, we decided to publish the proceedings in the languages in which the planning committee could write so as not to place added labor on conference attendees. These languages include Ayuuk/Mixe, Tu'un savi/Mixteco, Dixhzaa/Zapoteco, Spanish, and English.

As we designed this proceedings document, our team also had to make numerous decisions about how we would represent the various languages included. Rather than privilege Spanish or English, we chose to always foreground the Indigenous languages included in the document. We also learned that it would be helpful to label each language throughout the document so that readers who cannot understand certain languages will still know which languages are included in the documentation. For example, the title of the proceedings document, *Indigenous Language Interpreters and Translators: Toward the Full Enactment of All Language Rights*, was written with each language labeled in this way:

MIXE: Nëë ijty kyäjpnkëëxm ëts jä'ä ääw ayuuk nyajtëkmujkmëty mëtipe yäjk janyajxpën yäjk käjpxnäxpën

ZAP: Gal rxa'g za'a mniit ni rtuid de dixh xhte de gxh baañ

MIXT: tiun ni sa´a na saya´a tu´un saá si tu´un savi

ESP: Intérpretes y Traductores de Lenguas Indígenas: Hacia un Ejercicio Pleno de los Derechos Lingüísticos

ENG: Indigenous Language Interpreters and Translators: Toward the Full Enactment of All Language Rights

This same style of labeling is echoed throughout the document so readers can still see the presence of the various languages represented, even if they cannot read the entire document. Our goal in making these decisions was to break established hierarchies among languages and to de-center the emphasis on colonial languages (i.e., Spanish and English) that is often included in written publications. Taking feedback from participants who demonstrated that the Indigenous languages should be given space and attention in publications that discuss Indigenous communities, the goal of the multilingual content strategy developed for this unconference documentation protocol was to make rhetorical arguments about the value and importance of Indigenous language and Indigenous language speakers. While we certainly recognize that our decisions are imperfect, as it would have been ideal to have more languages represented in the document in various capacities, the process of designing this multilingual technical text required that unconference organizers consider how issues of power and positionality could be represented in written form. For this reason, in compiling the unconference proceedings, the unconference committee as a whole decided to incorporate stories and transcripts of dialogue that took place at the roundtables. We chose to create both a physical and a digital version of the proceedings so people who preferred or needed to access the content through sound could do so using edited audio recordings of the roundtable conversations. In this way, we wanted to use written documentation in the form of unconference proceedings to illustrate the "different modes of theorizing, speaking, writing, and making" fostered by Indigenous language speakers, interpreters, and translators (Torrez et al. 2019, 44).

## CO-DESIGNING TECHNOLOGIES TO SUPPORT INDIGENOUS SOVEREIGNTY

As our team designed the unconference proceedings, we also engaged in conversations about the role technologies play in the preservation and sustainability of Indigenous languages. At the conclusion of the unconference, all event participants gathered in a meeting hall as we discussed next steps for our ongoing collaborations. We designed an action plan that stemmed directly from their conversations on topics such as professional development for Indigenous language translators and interpreters, coalition building, accountability, and others. At the conclusion of the unconference event, many roundtable participants pointed to the power of technology as a potential avenue for staying connected and for continuing the work of advocating for the critical importance of Indigenous language interpretation across the world. These conversations directly countered the "dominant narrative" regarding Indigenous communities and technological innovation, a narrative that often positions Indigenous communities as technologically "illiterate" or resistant (Gómez Mentjívar and Chacón 2019, 3). As many scholars have noted, positivist notions of technology have historically (and contemporarily) led to the "extraction" (Gómez-Barris 2017) of resources, information, and land from Indigenous peoples. Yet as scholars also point out, Indigenous communities are consistently reclaiming technology as a tool for survival and activism as "the practice of indigenizing technology transforms new media into a tool for survival, and, more specifically, into a survivance story" (Gómez Mentjívar and Chacón 2019, 6).

At the unconference conclusion gathering, Indigenous language speakers echoed this push to reclaim technological innovation through Indigenous epistemologies that emphasize survivance. Moreover, unconference participants reflected on the multiple strategies they themselves have been using to leverage technology in their work, noting that as the unconference concluded, our newly formed community could continue to expand on these forms of communication to further highlight and advocate for Indigenous language sustainability and preservation.

For example, participant Shara spoke up at the conclusion gathering by reflecting on her work as an Indigenous language advocate who leverages both digital and analog technologies in her work. As Shara explained:

> Since 1986, I've been working with my legal radio broadcasting program, *Willakuyta*, which is broadcasted in Quechua. I've noticed the importance of having communication channels reach our Indigenous communities as quickly as possible, and I've also noticed that many Indigenous communities rely predominantly on radio channels to get information.
>
> But it's also important for us to use and develop other technologies. Communication channels are a force in the state, which is why we advocate for the visibility of our culture and languages in current communication channels, including radio, television, and the internet.

> Communication channels are spaces that we as Indigenous people should be working to take over progressively, rising and opening doors in order to be able to speak our languages through these mediums. We can even call these healing spaces because our community continues facing sociopolitical violence, and we need to create spaces to heal from this oppression.

In her work as an interpreter, Shara has also created a digitized glossary of common legal terms in Quechua, using this technology to, as she explained during the unconference closing session, "keep track of the legal terms that I was learning in order to decrease translation time. What used to take me six hours to translate now only takes three, meaning that I was able to reduce my translation time by 50 percent through the strategies that I developed day by day as an interpreter as well as through the use of this glossary."

Like Shara, many unconference participants embraced the use of technology and its role in Indigenous language translation, interpretation, and advocacy. From suggestions to create WhatsApp groups to stay connected as unconference participants, to the idea of building an online network of Indigenous language interpreters and translators, to the important idea of using digital technologies to develop and share best practices for successful translation and interpretation in Indigenous languages, participants at the unconference proposed innovative potential solutions for navigating the challenges and constraints Indigenous language interpreters and translators currently face in their work.

Yet many participants at this event also noted how technologies are consistently used to perpetuate violence against Indigenous people, particularly in the Mexican court system and in established infrastructures for Indigenous language translation and interpretation. For example, while there are infrastructures for Indigenous language interpreters to provide interpretation services through video conferencing in the US and other countries, there are no established payment systems for compensating interpreters for their work. Online dictionaries, with the exception of some like Shara's Quechua-Spanish legal glossary, are often built and supported by Western institutions that create systems that remain inaccessible to Indigenous language interpreters who live in rural communities with limited access to Wi-Fi and who communicate using non-Roman alphabets.

In her discussion at our closing roundtable, Erika Cuevas, a Chinateco speaker and director of Interpreters y Promotores Interculturales (Intercultural Interpreters and Promoters), explained:

> Technologies have always been used against us, but we need to reclaim them. Instead of the common stereotypes that we often see on social

media, we should use technology strategically by leveraging social media to share, for example, documentaries or short videos that contain information about public institutions that may be able to help us.

For example, I'm currently running the page for the Indigenous and Afroamerican Defense Directory at the Public Defender's Office, and there we are constantly uploading videos in different languages. We recently shared sixteen videos in different Indigenous languages spoken in the state, where we are sharing information about the services provided by the Public Defender's Office. We can use technology to make people aware of who we are and to make our community more aware of our rights.

Erika's emphasis on using technology to advocate for Indigenous language visibility and Indigenous people's rights points to an important opportunity for technical communicators to continue to develop multilingual technical communication projects that emphasize social justice. As Shara, Erika, and other unconference participants note, there is an opportunity to develop new tools, technologies, and systems to amplify, rather than extract or co-opt, Indigenous epistemologies and orientations to technological innovation. In thinking about and challenging my own work through the teachings of Indigenous language interpreters like Shara and Erika and others at CEPIADET, I'm pushed to think about technology's potential for Indigenous sovereignty and liberation. In my opinion, the most interesting and valuable part of the unconference attendees' focus on technology is not necessarily the use of digital technologies but instead the establishment of trust and collaboration that can lead these networks to truly be designed, adapted, and localized for Indigenous language activism. As Teotitlán del Valle Zapotec language activist Janet Chávez Santiago (2021) explained in her keynote presentation at the 2018 Digital Humanities Conference in Mexico City: "Digital media can be seen as a warp on which the speakers of Indigenous languages have an opportunity to weave their word and to share it within their own community and beyond. Although in our times digital media and social networks are a practical part of our daily lives and of our interactions with the world, we as speakers of Indigenous languages must truly appropriate these spaces, to weave our word well, in order to liberate ourselves" (i4).

As technical communication scholars and teachers continue to draw connections between multilingualism and technology, we should keep in mind that decolonial innovation requires an adherence to community goals, values, and histories that can too easily be ignored when language and technology are approached through settler colonial ideologies dominated by individualistic capitalist values (Haas 2012).

The Indigenous language interpreters I have the privilege to work with illustrate clearly and concretely for me how and why our policies and theoretical discussions of language diversity in technical communication should be directly tied to the broader goal of decolonization, one that can only move toward success through the active and continual push for sovereignty and land reclamation for and with Indigenous people.

## IMPLICATIONS FOR MULTILINGUAL TECHNICAL COMMUNICATION IN GLOBAL CONTEXTS

Throughout my collaboration with CEPIADET, I questioned (and continue to question) my role and the impact, both positive and negative, my presence could or should have on CEPIADET's work across communities. As Cana Itchuaqiyaq (2021) explains, when technical communicators work with marginalized communities whose backgrounds, histories, and languages differ from those of the researcher, embracing an Indigenist research paradigm informed by northern Alaskan Iñupiat values we should, among other things, ask "individuals to understand who they are, how they represent themselves and the world around them, and how they affect others" (46). For me, as a Bolivian researcher who does not speak an Indigenous language, this collaboration with CEPIADET pushed me to consider how my presence in certain conversations and activities inherently shifted the dynamics of the multilingual spaces in which CEPIADET consistently works. At the same time, as many technical communication researchers have documented and as members of CEPIADET and our research team communicated throughout this project, technical communication researchers can and do make important contributions to various communities and contexts through multilingual, transnational work, even when we are working with communities whose languages and histories are radically different from those we experience in US technical communication contexts.

Rather than trying to prevent a change in atmosphere when I was present at CEPIADET events and when information needed to be translated for me to understand a conversation with the participants represented in this study, this project pushed me to instead consider *how* my presence was shifting the space in relation to those around me and how I could encourage these changes to be helpful (rather than harmful) to CEPIADET's ultimate goals and mission. As global technical communication researchers who work in multilingual contexts, our presence will always take up space and change the dynamic of a research context. The key, I argue, is to recognize these shifts, acknowledge their existence, and work with participants to ensure that our impact yields positive results for our project collaborators. This means, for example, moving away from a presumption that the presence of an interpreter is not impacting our global technical communication research projects and instead leveraging our ability to connect with participants through interpreters and to value interpreters' influence as intellectual contributions to our work (Walton, Zraly, and Mugengana 2015). This can also mean recognizing the physical and emotional journeys interpreters, translators, and multilingual participants engage in when contributing to global technical communication projects and making space for participants to discuss and process these journeys, even when these discussions don't directly align with a specific project's goals or research questions. While the work of translators and interpreters in technical contexts is often intended to remain invisible, a rhetorically nuanced, justice-driven reframing of this orientation is long overdue and critically important, both in technical communication research and in professional practices more broadly.

More generally, the Indigenous language interpreters and translators included in this chapter can help technical communicators see that creating space for language diversity should not always focus on words alone but should instead constantly account for the relationships among language, land, and bodies that are always at play in multilingual communication. For example, as technical communicators continue to do valuable research in multilingual transnational contexts, we should strive to move, as Carlos Evia and Ashley Patriarca (2012) explain, "beyond compliance" in translating and interpreting our work. Rather than simply ensuring that a translator is present when we conduct research with multilingual communities, we can think about the material impacts this translation work has on the people we are working with, asking questions such as:

- How is the fact that my interviews are being translated influence the way my participants are responding?
- In what physical environment is this translation work taking place?
- Who is taking on the labor of language negotiation in these interactions, and how can I, as a researcher, contribute more to the work of language transformation in this context?
- Whose languages are taking up space in my interactions with multilingual communities? How am I asking questions?
- What answers am I expecting, either implicitly or explicitly, and (how) are these expectations reflective of my own colonial ideologies?

- What can I do to shift these expectations to open up room for more participation?

Of course, there are no perfect answers to these questions, as each response is context-dependent and flexible. At the same time, however, as the participants in this project demonstrate, questions like these are critical in working toward more equitable, respectful language relations, both in technical communication research and in broader interactions with multilingual communities. As technical communication researchers continue working to develop social justice–driven methods and methodologies, Indigenous language interpreters and translators can provide expertise through their extensive experience navigating linguistic and cultural differences to accomplish structural change for historically oppressed communities.

## CONCLUSION

The experiences shared by Indigenous language translators and interpreters in this chapter helped me understand that multilingual technical communication experiences extend far beyond words, instead encompassing relationships among words, people, cultures, tools, technologies, and lands. In multilingual technical communication research, we often focus primarily on finished products—on translated interfaces, multilingual technical texts, and multilingual media that can be shared and distributed in global contexts. While these final products of multilingual technical communication are certainly important, in proposing that technical communicators think critically about designing multilingual *experiences,* I emphasize the importance of considering all elements that go into multilingual design—including especially the people who make these designs possible and the journeys these people undertake consistently to provide language access for and with their communities. For Indigenous language translators and interpreters such as those depicted in this chapter, multilingual technical communication does not start or end with a successful interpretation and translation project, especially when translation and interpretation take place on colonized lands through Western colonial systems. While it would be impractical to propose or assume that technical communicators can single-handedly decolonize lands and completely subvert colonial systems, what we can do is what we have always done as a field: make technical processes visible and understandable to a wide range of stakeholders, facilitate communication across communities and contexts, and, by embracing a social justice perspective to technical communication praxis, build reciprocal partnerships with community members so we can contribute to, rather than prevent or harm, Indigenous sovereignty and survivance.

Considering the lessons the Indigenous language interpreters and translators shared in this chapter, in chapter 7 I move on to extend broader implications for multilingual technical communication research. These implications derive from all the participants who have been kind enough to share their brilliance throughout this book, including the bilingual youth and communities in El Paso and Ciudad Juárez, students and community members at SAFAR in Kathmandu, and CEPIADET's collaborative unconference planning team. Through a discussion of the multiple implications each set of participants has presented for global technical communication research, I will provide practical strategies contemporary technical communicators can apply to their work. I emphasize that although our research practices will never be perfect, through an intersectional, interdependent orientation, technical communicators can continue to centralize the value of linguistic, racial, and cultural diversity in the field—acknowledging the multilayered impacts linguistic transitions have in our projects, in our classrooms, and in all the spaces we inhabit.

Before moving on to the concluding chapter, in the section that follows I include a reflection on this chapter and on the unconference event written by Erika Cuevas. This reflection was written in 2021, two years after the unconference event, and it outlines Erika's current work and our ongoing collaborations.

## REFLEXIÓN DE ERIKA CUEVAS

En mi estado se hablan 16 lenguas indígenas, de las cuales derivan 177 variantes y entre ellas se encuentra el chinateco alto de San Pedro Yolox, el cual orgullosamente hablado desde mi niñez. El chinateco lo aprendí a hablar gracias a mis padres; quienes a pesar de que escuchaban de mis abuelos decir, que lo único que debían enseñarles a sus hijos era el español, porque si salíamos de nuestro pueblo la gente de la ciudad nos iba a discriminar por hablar diferente.

Aunque mis padres hubieran querido que sólo hablara el español, fue inevitable aprender la lengua, ya que en la escuela ò cuando jugaba con mis primos y hermanos, todo el tiempo hablábamos en chinanteco.

In my state [Oaxaca], more than 16 Indigenous languages are spoken, from which stem over 177 variants. Among these languages is Chinateco alto de San Pedro Yolox, which I've proudly spoken since childhood. I learned to speak Chinateco thanks to my parents, despite the fact that they were consistently told by my grandparents that they should only teach their kids to speak Spanish, since we would be discriminated against outside of our town for speaking a different language. Although my parents may have preferred that we only learn to speak Spanish, learning our language was inevitable, since in school or when playing with my siblings and cousins, we would always speak in Chinateco.

Pasó el tiempo, y decidí salir de mi pueblo. Me fui a estudiar la secundaria y el bachillerato. Fue complicado porque en dónde vivía no tenía con quien hablar Chinanteco, ya que todos hablaban español, y a veces hablaban de cosas que yo no sabía, o bueno, no entendía sus expresiones, pero afortunadamente jamás sufrí de discriminación o bullying en la escuela por mi origen, al contrario, recuerdo que cuando entre a la universidad, mis compañeros al escucharme hablar en mi lengua me decían que querían aprenderlo, así, en mis tiempos libres les enseñaba ciertas palabras y expresiones en mi lengua.

Time passed, and I decided to leave my community. I went to high school and to get my bachelor's degree. This was complicated, since where I lived I didn't have anyone with whom I could speak Chinateco, since everyone spoke Spanish. Sometimes they would talk about things that I didn't know or, well, I didn't understand their expressions, but fortunately I never suffered from discrimination or bullying in school due to my origin. Instead, I remember that when I started college, my classmates heard me speaking my language and said that they wanted to learn it, so in my free time I would teach them certain words and expressions in my language.

Paso el tiempo y en mi último año de la carrera tenía que realizar mi servicio social, mentiría si les cuento que estuve buscando una institución que estuviera relacionado con el tema de cultura o revitalización de lenguas indígenas, pero la vida es muy rara que termina colocándote en el lugar que te corresponde, y fue así como me encontré con la Defensoría Pública de Oaxaca, donde conocí a Eduardo Martínez Gutiérrez, quien es una persona que me inspirara día a día al ver su entrega en el trabajo de revitalizar las lenguas indígenas y con quien hemos trabajado en un proyecto llamado Intérpretes y Promotores Interculturales.

Time went on, and in my last year of college I needed to do community service. I'd be lying if I said that I was looking for an organization related to culture or Indigenous language revitalization, but life is strange and ends up placing you where you belong. This is how I found the Public Defender's Office of Oaxaca, where I met Eduardo Martínez Gutiérrez, a person who inspires me daily as I witness his work in Indigenous language revitalization and with whom I have worked on a project called Intercultural Interpreters and Promoters.

El proyecto surgió con el objetivo de garantizar los derechos lingüísticos y procesales de las personas indígenas privadas de libertad y evitar retrasos en los expedientes por ausencia de Intérpretes, un proyecto en el que participan más de 90 jóvenes hablantes de las diferentes lenguas indígenas. El proyecto va más allá de auxiliar un proceso penal, el equipo de Intérpretes y Promotores interculturales se convirtió en una familia, se convirtió en un espacio que nos permite reencontramos con nuestra identidad, con nuestra cultura, se convirtió en aquel respaldo que muchos necesitábamos para seguir promoviendo y fomentado nuestra lengua.

This project was established with the goal of guaranteeing the linguistic and procedural rights of Indigenous people, protecting their rights, and avoiding delays in their court proceedings due to a lack of interpreters. More than ninety young people who speak Indigenous languages participate in this project. This project provides more than legal assistance during legal proceedings, since the team of Intercultural Interpreters and Promoters has become a family, a space that allows us to rediscover our own identity and culture and a space that has become the support many of us needed to continue promoting and growing our language.

Nuestro equipo ha asistido a más de 500 interpretaciones en el estado y también en tribunales internacionales, hemos participado en diferentes campañas audiovisuales para promover los derechos humanos, nos hemos convertido en el puente de comunicación de nuestras comunidades con los diferentes servicios públicos. Pero no solo hacemos eso, cada uno de nosotros estamos comprometidos en realizar acciones en nuestras comunidades para preservar nuestra lengua, aunque no es tan fácil como se escucha, realmente es una tarea muy difícil, no siempre contamos con los medios necesarios para hacerlo, en mi caso, cada que viajo a mi pueblo me reúno con los niños para jugar con loterías, memoramas y otro juegos de mesa con la finalidad de que el tiempo que paso con ellos, lo usemos para fomentar el uso y aprendizaje del Chinanteco.

Our team has facilitated more than 500 interpretations in the state and in international courts. We've also participated in various audiovisual campaigns to promote human rights, and we have become the communication channel between our communities and different public services. Furthermore, each one of us is committed to carrying out actions in our communities to preserve our languages, although this isn't as easy as it sounds. It's a difficult task, since we don't always have the necessary means to carry out these goals. In my case, every time I travel to my community, I get together with the kids to play lotería, memory, or other table games so we spend our time together practicing and expanding the learning of Chinateco.

Considerando el panorama actual, en donde las tecnologías desempeñan un importante papel en nuestra vida, trabajo en fomentar el aprendizaje y uso del Chinanteco a través de audio-cuentos que son de mi autoría. He tenido la intención de traducir caricaturas animadas y películas, sin embrago, para hacerlo, necesito los permisos de Disney y de Warner Bros, y contactarlos nos ha sido imposible, pero seguiremos intentado y buscando otros medios para promover nuestra lengua, con tal de no dejarla morir, ya que si nuestra lengua muere, morirá nuestra esencia e historia.

Considering the current landscape, where technologies play an important role in our lives, I also work to foster the learning of Chinateco through audio-stories that I write myself. I have the goal of translating animated cartoons and movies. However, to do this, we need permission from Disney and Warner Bros., and contacting them has been impossible. However, we will keep trying and finding other ways to promote our language to prevent it from dying, since if our language dies, our essence and history would die as well.

Hoy, después de toda esta experiencia, puedo decir que contrario a lo que pensaban mis abuelos sobre el Chinanteco como una limitante para mi futuro y mi desarrollo profesional, el Chinnateco se ha convertido en la llave que me ha abierto muchas puertas para intercambiar experiencias. No tengo palabras para describir la satisfacción que siento al ver que fue todo lo contrario, que gracias a mi lengua se me abrieron puertas laborales, he conocido muchas culturas y también me ha permitido encontrarme con personas maravillosas, como lo es la autora de este libro [Laura Gonzales].

Today, after all of these experiences, I can say that contrary to what my grandparents thought about Chinateco as a limitation for my future and my professional development, Chinateco has become the key that has opened many doors to exchange experiences. I don't have the words to describe the satisfaction that comes with seeing that thanks to my language, many professional doors have opened for me, I've learned about many different cultures, and I've been able to meet wonderful people, like the author of this book [Laura Gonzales].

# 7

# IMPLICATIONS FOR DESIGNING MULTILINGUAL EXPERIENCES IN TECHNICAL COMMUNICATION

There is a long history of technical communication researchers going into community spaces, in both local and global contexts, to conduct research and work with community members to design tools and technologies for various purposes. When reading this research, as well as when planning my own collaborative projects such as the ones I describe in this book, I frequently ask myself: what happens when we (researchers) leave? If designing and collaborating with multilingual communities is a multilingual experience grounded in trust and relationship building, what happens to that trust and to those relationships at the conclusion of a project? In other words, what aspects of these multilingual experiences remain sustainable?

Indigenous scholars have long argued for the importance of relationality in research in collaboration. For example, Angela M. Haas (2015) explains that as humans and teachers, we "construct ourselves and O/others in relation to one another, to visuals and visibility, and to technologies" (206). In this book, I've tried to demonstrate that this type of relationality, how we "construct ourselves and O/others in relation to one another," is often mediated through language—through the verbalized and the silent, the material and the embodied, the alphabetic and the multimodal elements through which humans communicate their thoughts, ideas, and feelings (Haas 2015, 206). The relationships that come together in multilingual experiences have no end date, whether our research protocols recognize this or not. In each of the projects we take on as technical communicators working in multilingual, global contexts, we impact the communities that trust us with their time, energy, and labor; and this impact extends beyond the parameters set by Western notions of time and space.

At the core, I hope the stories presented in this book can provide technical communication researchers and practitioners who work in community contexts with different lenses through which we can

https://doi.org/10.7330/9781646422760.c007

continue to study language diversity and the role it plays in our research. By pairing discussions of localized projects in the field of technical communication with various interdisciplinary theoretical and methodological frameworks, I hope to demonstrate how technical communication researchers can continue to acknowledge that our work as communicators is never neutral and that the impact we have on our surrounding communities, environments, and contexts should not be ignored. As Miriam F. Williams (2013) explains, "Those of us who examine historical representations of race and ethnicity in technical communication will not have to do so without excellent examples of rhetorical, ethical, and historical analyses," since we can rely on emerging social justice–driven research in our field as well as on "theoretical perspectives and methods from other areas that many of us [in technical communication] have never studied or are unfamiliar with, but that are used by scholars in ethnic studies, sociology, and other disciplines" (89).

While it is difficult to generalize implications or neat lists of "practitioner takeaways" from localized projects stemming from different contexts and communities across several years, in this chapter I highlight potential implications technical communication researchers can extend from the projects presented in this book so that we as practitioners in this field can collectively continue working to design multilingual experiences in technical communication that honor the various languages and communities that fuel and sustain our work.

## ENGAGING THE COMPLEXITIES OF LANGUAGE ACCESS

In each of the projects presented in the previous chapters, participants of all ages were critically aware of the role language plays in providing or limiting access to their knowledges and ways of learning. At La Escuelita, for example, youth chose which parts of their website to translate for particular reasons and particular audiences. At the unconference event coordinated by CEPIADET, participants chose which languages to use in specific situations, all with a critical understanding of who would be included and excluded in the discussion depending on individual rhetorical linguistic choices. Through an intersectional, interdependent methodology, technical communication researchers can acknowledge participants' linguistic choices not as barriers to be overcome—for example, participants are not speaking in English because they "don't know" how to communicate something in that language—but as opportunities to recognize how participants' rhetorical linguistic choices are shaped by other factors in communities' lived experiences, which are often influenced by racism and colonization. Thus, as technical communication researchers seeking to embrace and sustain rather than co-opt or dominate multilingual experiences with our participants, it's important to recognize that sometimes language accessibility is not the ultimate goal of an interaction. Instead, if we as technical communication researchers are not able to engage in our participants' heritage language(s), we can work to develop alternate methods of listening, to what is both said and unsaid, in our research contexts. As innovative information designers, intersectional and interdependent approaches to multilingual experiences can help technical communicators recognize how communicative spaces have multiple access points, some of which do not and should not rely on words in a shared language.

Furthermore, in situations where technical communication researchers are communicating with participants in a shared language, as was the case in the English communication that took place with my participants in Nepal, we should not assume that the availability of translation or the presence of a shared language means that communication is actually accessible. Instead, we should consider how the various positionalities of the languages we use in any research space influence who can actually participate in a discussion—who will be motivated to share and who will choose to remain quiet due to various issues and lived experiences. As the Indigenous language interpreters and translators from CEPIADET explained, the presence of language diversity in colonial contexts can still uphold colonial values; thus, having the opportunity to communicate with participants in the language that is most comfortable to us as technical communication researchers (e.g., English) may not actually create a participatory, accessible space where all participants feel welcomed and can engage. It is up to us as researchers, in our responsibility within the design of multilingual experiences, to research the language relationships of the contexts in which we work, to interrogate how language positionalities as well as our own embodied presence are impacting or contributing to colonial legacies, and to take on the burden of language transformation rather than leaving this burden in the hands of the participants who are already taking on the labor of our work—often without compensation.

## USING LANGUAGE TO MAKE SPACE FOR PARTICIPATION

At the unconference in Oaxaca described in chapter 6, I explain how Indigenous language interpreters and translators made it a point to allow their languages to take up space in our conversations, even when nobody except the speaker could understand the words spoken in a particular language. In these situations, Indigenous language interpreters and translators demonstrated how language is tied to power and land and how it is important to speak our languages in order to represent who we are as human beings and to honor our communities and ancestors. This notion of allowing language to take up space can also inform how we as technical communication researchers embrace our work with multilingual communities as we de-center the dominance of standardized English and Western communicative practices.

Often, in multilingual technical communication projects, researchers may want to minimize the time or space language "barriers" play in the course of a project. Wanting to get to the "point" or the "core" of the work, we can choose to outsource language work or to ignore the presence of translation if our participants speak English (even when English is not a heritage language in the community). When given the option to rely on English, technical communication researchers may proceed with the project without acknowledging the translation work our participants are doing, often silently, to contribute knowledge to our research.

Intentionally crafting and sustaining multilingual experiences in technical communication pushes technical communication researchers to move beyond a reliance on any individual language, working instead to allow language transformation to take up space in the research process. Methods such as participatory translation, as demonstrated in chapters 4–6, can encourage multilingual communities to recognize their own expertise in translation and thus feel more comfortable contributing to a collaborative project. At the same time, participatory translation is not a quick fix for power inequities in participatory projects and should be incorporated in multilingual research only as one avenue through which participants can make contributions. While many multilingual communities, and multilingual communities of color in particular, have been made to feel as though they are "not good" with technology and thus have nothing to contribute to technology design, an emphasis on the importance of translation and multilingualism in contemporary technological innovation can open up avenues for discussion by and intervention from multilingual community members. This requires that technical communication researchers allow language to take up space—as participants debate possible translations; share stories related to their own interpretations of words, concepts, and ideas; and use their own linguistic practices to shape communication. Moving beyond technical communication's long-standing emphasis on expediency and efficiency (Jones 2016)—concepts grounded in white/Western approaches to communication—is critical to engaging in multilingual experiences that honor community knowledge and labor.

## RECOGNIZING THAT LANGUAGE IS MORE THAN WORDS

The disability studies concept of the bodymind, as described by Margaret Price (2011) and Sami Schalk (2018), pushes researchers to recognize that the body and the mind are not separate entities but are instead interconnected and interdependent. Disability studies researchers such as Christina V. Cedillo (2018) also point to the multiple elements embodied in disability, some of which are visible and some of which remain invisible to and undiagnosed by Western medicine. This recognition of the bodymind and of the connections between visible and invisible embodiment is critical to an understanding of multilingual experiences as those that encompass much more than words. The translation of information from one named language to another can foster language access, but, as the participants in this book make clear, language access that focuses only on translation can still be oppressive and violent, particularly for multilingual communities of color whose values and histories are consistently excluded and colonized. When researchers ask participants to communicate in a language like English, which may differ from participants' heritage language or preferred modes of communication, we may be triggering memories and embodied experiences that often go unrecognized and ignored in multilingual projects. The process of translating, particularly into colonial languages, can encompass invisible trauma and erasure that have long been used to oppress multilingual communities. While we may not be able to avoid using colonial languages in research with multilingual communities, an intersectional, interdependent orientation to this work requires that researchers recognize the invisible aspects of translation and their role in the work we do with communities across contexts. To translate is to engage in collective transformation of ideas, histories, experiences, thoughts, and feelings—all of which can be loaded with multiple elements that are often unrecognized in formal research settings. As this book's participants illustrate, language is a fluid, malleable, multidimensional practice that consistently impacts our bodyminds.

I also want to emphasize that my goal in threading disability studies concepts like the notion of the bodymind and the concept of interdependency in conversations about translation and language access was to bring more attention to the way disability studies scholars have long been considering issues of language accessibility in disabled communities. While the participants highlighted in this book were not asked to disclose specific disabilities, designing multilingual experiences with disability in mind can result in more accessible research spaces that foster participation by multilingual communities of color. As future researchers continue to extend the concepts in this book in other global technical communication research, I hope we can have more extended conversations about the interconnected nature of language and disability studies research to continue honoring the expansive intersectional identities of global communities.

## DESIGNING LANGUAGE BEYOND BINARIES

As demonstrated in the discussion of language on the border represented in chapter 4, language, like people, should not be reduced to binaries or categorized into "this" language, culture, or label "or" another. Instead, as technical communication researchers continue to design multilingual experiences alongside communities, we should develop frameworks for embracing complexity in communicative practice. These frameworks, in conjunction with the expertise of multilingual communities, can facilitate the design and development of more effective multilingual technologies that reflect the fluidity of language rather than segmenting languages into static categories.

At the unconference event with CEPIADET, for example, working within standardized language categories would have erased the extensive linguistic, cultural, and racial diversity embedded in our collaborative space. By choosing to limit communication to the language(s) that are shared by all members of a group (e.g., Spanish), designers may erase the presence of other communicative practices that are important to users and their communities. Likewise, when designing digital tools and interfaces, designers often have to choose the specific language(s) through which their users can interact on a specific aspect of the platform. Sometimes, organizations that serve multilingual communities have to make difficult decisions about the language(s) they will use in their materials to reach (or not) their target users or audiences (Cardinal 2019). An intersectional, interdependent approach to multilingual technical communication, in this case, can help designers see that not all content in every part of an interface needs to be in the same language and that not all content needs to be translated into every language. Instead, as Margaret Price and Stephanie L. Kerschbaum (2016) explain in their discussion of (in)accessibility in research methodologies, when designing in multiple languages, designers are always making access choices that will impact various stakeholders in different ways. Rather than aim for complete accessibility (which is impossible and can be harmful), working to design multilingual experiences rather than perfectly translated interfaces can push designers to find ways of embracing the fluidity of language to develop tools and technologies that don't follow standardized translation practices. As the youth at La Escuelita demonstrate, different audiences will leverage their linguistic abilities to engage with multilingual content in their own ways, and the role of the designer is to facilitate these interactions while also recognizing that all content does not need to be made accessible for every person in the same way at every juncture. Visuals, colors, sounds, and words are malleable and fluid and can help users understand and engage beyond set or assumed linguistic boundaries.

## LIMITATIONS AND FUTURE DIRECTIONS

In sharing the stories of how I co-designed multilingual experiences across contexts, my goal is to illustrate the multiple complexities embedded in multilingual technical communication. Rather than present a clean or straightforward list of dos and don'ts for working with multilingual communities in global contexts, my goal in this book is to open up and expand conversations about how technical communication researchers can further consider the role language plays in all our work—whether we are working in an office, in a classroom, or in a community context. I'm aware that the stories in this book may not resonate with everyone, as they reflect grounded experiences in very specific contexts. However, my hope is that by reading and gaining some insights into the multilingual experiences represented here, technical communication researchers, students, and practitioners can expand the lessons presented into other contexts, working with different communities that speak different languages in different contexts. No single book can encompass work in all languages, and there is no list of tips that would be useful across all communities. Thus, there are undoubtedly limitations to the languages represented here and, even more so, to the specific methods and actions I took in these communities. While my goal was to embody an intersectional and interdependent methodology that fosters justice rather than causes harm, there is no guarantee that I did so successfully, as all of the relationships represented here are still ongoing and developing. In the spirit of sharing both successes and failures, the work included in this book is a call to action to continue to recognize the power of language in global technical communication research. I look forward to seeing if and how the frameworks I presented in these projects can be expanded by others.

## CONCLUSION

Through the projects depicted in this book, I make a case for reframing language difference in technical communication, from design problem to design opportunity (Cardinal, Gonzales, and Rose 2020). At the same time, I argue that multilingual communities of color in particular are consistently harmed by white/Western researchers who seek to extract (Cardinal, Gonzales, and Rose 2020) knowledge from communities while also upholding long-standing assumptions about multilingual communicators' "lack" of communicative competence and intellect.

In thinking about the implications of this book within technical communication's social justice agenda, I think it's important to highlight that the multilingual communities represented here can help technical communicators continue to expand the field's notion of what "effective" communication should look like. In a recent article detailing "the most bothersome errors" technical communication employers identify in prospective employees' writing, for example, Carolyn Gubala, Kara Larson, and Lisa Melonçon (2020) argue that contemporary technical communication employers are "bothered" by "errors" in technical communication contexts. These authors advocate for "the importance of error-free writing," noting that technical communication employers viewed professionals who had "errors" in their writing as "poorly educated" as well as "hasty, careless, uncaring, or uninformed" (269). When put in the context of the projects described in this book, we can see how long-standing assumptions and presumptions about the way language works result in categorizations of people as "hasty" and "careless" when they might, in fact, simply be like the borderland residents such as those who contributed to the design of the Diabetes Garage brochure. When we perpetuate language ideologies that uphold white/Western/standardized and colonial values, we are causing community members like those represented in chapter 4 to feel as though their hybrid language practices are errors rather than reflections of effective, nuanced, and expert contemporary communication.

When technical communicators embrace binary, static notions of language and when they position language and translation as separate from the work of technical communication practice, they run the risk of embracing oppressive perspectives that extend racist, colonial legacies consistently imposed on multilingual communities. When we view language as "correct" or "incorrect," as "accurate" or "error," we assume what Natasha N. Jones and Miriam F. Williams (2018) warn against—the neutrality of technical documents that can appear benevolent while instead functioning as "technologies of disenfranschisement" that perpetuate racist and ableist values (371).

Embracing multilingual experiences as part of technical communication's efforts toward social justice requires a realignment of how the field views language—how we perceive effective communication, how we envision our users and their linguistic practices, and how we design information for the current global majority. While the projects depicted in this book only introduce brief examples of multilingual experiences in context, the participants who share their expertise in these projects demonstrate a broader need for further engagement with community knowledge in technical communication. Just as researchers such as Rebecca Walton, Kristen Moore, and Natasha Jones (2019), Jones and Williams (2018), and Angela M. Haas and Michelle F. Eble (2018) encourage technical communication researchers to move away from assumptions about technical communication's neutrality, the participants depicted in this book demonstrate both the consequences of these assumptions for communities of color and the multiple possibilities that are available to technical communication researchers if we take the time to make space for translation, embrace language fluidity, and recognize that our field's long-standing values of expediency and simplicity need to consistently be challenged if we want our designs to reflect the complexity of our contemporary world.

To design multilingual experiences in technical communication is to move beyond binaries, to center positionality, to pause and embrace complexity, and to reposition "error" and "messiness" as intentional elements of all language work. As technical communicators continue to engage in multilingual communication projects, we should remember that language, like technology, is never neutral and that the rhetorical choices we make in the languages we use when communicating with our participants have just as much, if not more, influence on the results of our work as do the research protocols, methods, and frameworks we embrace. As the participants in this book illustrate, language in its many dimensions mediates all communication and should thus be centralized rather than ignored in contemporary technical communication work. Through the stories of my participants and the relationships that are illustrated in this project, I hope technical communication researchers will continue to take up the responsibility of language transformation when working in community contexts, recognizing that access is never guaranteed and that power should never be ignored in discussions of language difference.

# REFERENCES

Acharya, Keshab R. 2018. "Usability for User Empowerment: Promoting Social Justice and Human Rights through Localized UX Design." In *Proceedings of the 36th ACM International Conference on the Design of Communication*, 1–7. New York: Association for Computing Machinery.

Agboka, Godwin Y. 2013. "Participatory Localization: A Social Justice Approach to Navigating Unenfranchised/Disenfranchised Cultural Sites." *Technical Communication Quarterly* 22 (1): 28–49.

Agboka, Godwin Y. 2014. "Decolonial Methodologies: Social Justice Perspectives in Intercultural Technical Communication Research." *Journal of Technical Writing and Communication* 44 (3): 297–327.

Agboka, Godwin Y. 2018. "Indigenous Contexts, New Questions: Interrogating Human Rights Perspectives in Technical Communication." In *Key Theoretical Frameworks: Teaching Technical Communication in the Twenty-First Century*, edited by Angela M. Haas and Michelle F. Eble, 114–37. Boulder: University Press of Colorado.

Agboka, Godwin Y., and Natalia Matveeva, eds. 2018. *Citizenship and Advocacy in Technical Communication: Scholarly and Pedagogical Perspectives*. New York: Routledge.

Alim, Sami H., and Alastair Pennycook. 2007. "Glocal Linguistic Flows: Hip-Hop Culture(s), Identities, and the Politics of Language Education." *Journal of Language, Identity, and Education* 6 (2): 89–100.

Alim, Sami H., John R. Rickford, and Arnetha F. Ball. 2016. "Introducing Raciolinguistics." In *Raciolinguistics: How Language Shapes Our Ideas about Race*, edited by H. Samy Alim, John R. Rickford, and Arnetha F. Ball, 1–30. New York: Oxford University Press.

Als, Benedikte S., Janne J. Jensen, and Mikael B. Skov. 2005. "Comparison of Think-Aloud and Constructive Interaction in Usability Testing with Children." In *Proceedings of the 2005 Conference on Interaction Design and Children*, 9–16. https://doi.org/10.1145/1109540.1109542.

Anzaldúa, Gloria. 1987. *Borderlands: La Frontera*. Vol. 3. San Francisco: Aunt Lute.

Baker-Bell, April. 2020. "Dismantling Anti-Black Linguistic Racism in English Language Arts Classrooms: Toward an Anti-Racist Black Language Pedagogy." *Theory into Practice* 59 (1): 8–21.

Banks, Adam J. 2006. *Race, Rhetoric, and Technology: Searching for Higher Ground*. New York: Routledge.

Batova, Tatiana. 2010. "Writing for the Participants of International Clinical Trials: Law, Ethics, and Culture." *Technical Communication* 57 (3): 266–81.

Batova, Tatiana. 2018. "Negotiating Multilingual Quality in Component Content-Management Environments." *IEEE Transactions on Professional Communication* 61 (1): 77–100.

Batova, Tatiana, and Dave Clark. 2015. "The Complexities of Globalized Content Management." *Journal of Business and Technical Communication* 29 (2): 221–35.

Benjamin, Ruha. 2019. *Race after Technology: Abolitionist Tools for the New Jim Code*. Oxford: Oxford University Press.

Berne, Patricia, Aurora Levins Morales, David Langstaff, and Sins Invalid. 2018. "Ten Principles of Disability Justice." *WSQ: Women's Studies Quarterly* 46 (1): 227–30.

https://doi.org/10.7330/9781646422760.c008

Bista, Krishna, Shyam Sharma, and Rosalind Latiner Raby. 2019. "Telling Stories, Generating Perspectives: Local-Global Dynamics in Nepalese Higher Education." In *Higher Education in Nepal: Policies and Perspectives*, edited by Krishna Bista, Shyan Sharma, and Rosalind Latiner Raby, 135–38. New York: Routledge.

Bloom-Pojar, Rachel. 2018. *Translanguaging Outside the Academy: Negotiating Rhetoric and Healthcare in the Spanish Caribbean.* Urbana, IL: National Council of Teachers of English.

Bloom-Pojar, Rachel, and Danielle DeVasto. 2019. "Visualizing Translation Spaces for Cross-Cultural Health Communication." *Present Tense: A Journal of Rhetoric in Society* 3 (7). https://www.presenttensejournal.org/volume-7/visualizing-translation-spaces-for-cross-cultural/.

Bow, Leslie. 2011. *Betrayal and Other Acts of Subversion: Feminism, Sexual Politics, Asian American Women's Literature.* Princeton, NJ: Princeton University Press.

brown, adrienne m. 2017. *Emergent Strategy: Shaping Change, Changing Worlds.* Chico, CA: AK Press.

Brown, Ariana. 2021. *We Are Owed.* Cleveland: Grieveland.

Butler, Janine. 2017. "Bodies in Composition: Teaching Writing through Kinesthetic Performance." *Composition Studies* 45 (2): 73–90.

Cardinal, Alison. 2019. "Participatory Video: An Apparatus for Ethically Researching Literacy, Power, and Embodiment." *Computers and Composition* 53: 34–46.

Cardinal, Alison, Laura Gonzales, and Emma Rose. 2020. "Language as Participation: Multilingual User Experience Design." In *Proceedings of the 38th ACM International Conference on the Design of Communication*, edited by Josephine Walwema, Daniel Hocutt, and Stacey Pigg, 1–7. New York: Association for Computing Machinery.

Cedillo, Christina V. 2018. "What Does It Mean to Move? Race, Disability, and Critical Embodiment Pedagogy." *Composition Forum* 39. http://compositionforum.com/issue/39/to-move.php.

CEPIADET AC. 2020, April 1. *Principal.* Oaxaca de Juárez, Mexico: Centro Profesional Indígena de Asesoría, Defensa, y Traducción A. C. www.cepiadet.org/about.html.

Chavan, Apala Lahiri. 2005. "Another Culture, Another Method." 2005. *Proceedings of the 11th International Conference on Human-Computer Interaction* 21 (2): 1–4.

Chávez Santiago, Janet. 2021. "Keynote: Tramando la palabra/Weaving the Word." *Digital Scholarship in the Humanities* 36 (1): i4–i8.

Clemmensen, Torkil. 2011. "Templates for Cross-Cultural and Culturally Specific Usability Testing: Results from Field Studies and Ethnographic Interviewing in Three Countries." *International Journal of Human-Computer Interaction* 27 (7): 634–69.

Collins, Patricia Hill. 1999. *Black Feminist Thought: Knowledge, Consciousness, and the Politics of Empowerment.* New York: Routledge.

Collins, Patricia Hill. 2019. *Intersectionality as Critical Social Theory.* Durham, NC: Duke University Press.

Combahee River Collective. 1977. "The Combahee River Collective Statement." https://americanstudies.yale.edu/sites/default/files/files/Keyword%20Coalition_Readings.pdf.

Concha, Jeannie Belinda. 2018. "Approaches to Enhancing Patient-Centered Communication in Caring for Hispanic/Latino Patients with Diabetes." *Journal of Clinical Outcomes Management* 25 (3): 122–31.

Constitución Política de los Estados Unidos Mexicanos. 2017. "Artículo 133." Mexico City: Porrúa.

Costanza-Chock, Sasha. 2020. *Design Justice: Community-Led Practices to Build the Worlds We Need.* Cambridge, MA: MIT Press.

Cox, Matthew. 2018. "Shifting Grounds as the New Status Quo: Examining Queer Theoretical Approaches to Diversity and Taxonomy in the Technical Communication Classroom." In *Key Theoretical Frameworks: Teaching Technical Communication in the Twenty-First*

*Century*, edited by Angela M. Haas and Michelle F. Eble, 287–303. Boulder: University Press of Colorado.

Crenshaw, Kimberlé W. 1989. "Demarginalizing the Intersection of Race and Sex: A Black Feminist Critique of Antidiscrimination Doctrine, Feminist Theory and Antiracist Politics." *University of Chicago Legal Forum* 8: 139–67.

Crenshaw, Kimberlé W. 2017. "Kimberlé Crenshaw on Intersectionality, More than Two Decades Later." *Columbia Law School*. https://www.law.columbia.edu/news/archive/kimberle-crenshaw-intersectionality-more-two-decades-later.

Cristancho, Sergio, D. Marcela Garces, Karen E. Peters, and Benjamin C. Mueller. 2008. "Listening to Rural Hispanic Immigrants in the Midwest: A Community-Based Participatory Assessment of Major Barriers to Health Care Access and Use." *Qualitative Health Research* 18 (5): 633–46.

Cusicanqui, Silvia Rivera. 2010. *Ch'ixinakax Utxiwa: On Practices and Discourses of Decolonization*. Buenos Aires: Tina Limón Ediciones.

de la Piedra, María Teresa, Blanca Araujo, and Alberto Esquinca. 2018. *Educating across Borders: The Case of a Dual Language Program on the US-Mexico Border*. Tucson: University of Arizona Press.

de los Ríos, Cati V. 2018. "Toward a Corridista Consciousness: Learning from One Transnational Youth's Critical Reading, Writing, and Performance of Mexican Corridos." *Reading Research Quarterly* 53 (4): 455–71.

Del Hierro, Marcos. 2018. "Stayin' on Our Grind: What Hiphop Pedagogies Offer to Technical Writing." In *Key Theoretical Frameworks: Teaching Technical Communication in the Twenty-First Century*, edited by Angela M. Haas and Michelle F. Eble, 163–84. Boulder: University Press of Colorado.

Del Hierro, Victor. 2019. "DJs, Playlists, and Community: Imagining Communication Design through Hip Hop." *Communication Design Quarterly* 7 (2): 28–39.

Del Hierro, Victor, Valente Francisco Saenz, Laura Gonzales, Lucía Durá, and William Medina-Jerez. 2019. "Nutrition, Health, and Wellness at La Escuelita: A Community-Driven Effort toward Food and Environmental Justice." *Community Literacy Journal* 14 (1): 26–43.

Ding, Huiling. 2020. "Crowdsourcing, Social Media, and Intercultural Communication about Zika: Use Contextualized Research to Bridge the Global Digital Divide in Global Health Intervention." *Journal of Technical Writing and Communication* 50 (2): 141–66.

Dorpenyo, Isidore. 2020. *User Localization Strategies in the Face of Technological Breakdown*. New York: Palgrave Macmillan.

Dorpenyo, Isidore, and Godwin Agboka. 2018. "Election Technologies, Technical Communication, and Civic Engagement." *Technical Communication* 65 (4): 349–52.

Driskill, Qwo-Li. 2015. "Decolonial Skillshares." In *Survivance, Sovereignty, and Story: Teaching American Indian Rhetorics*, edited by Lisa King, Rose Gubele, and Joyce Rain Anderson, 57–78. Boulder: University Press of Colorado.

Durá, Lucía. 2016. "What's Wrong Here? What's Right Here? Introducing the Positive Deviance Approach to Community-Based Work." *Connexions International Professional Communication Journal* 4 (1): 57–89.

Durá, Lucía, Laura Gonzales, and Guillermina Solis. 2019. "Creating a Bilingual, Localized Glossary for End-of-Life Decision-Making in Borderland Communities." In *Proceedings of the 37th ACM International Conference on the Design of Communication*, edited by Julie Staggers, Daniel P. Richards, Tim Amidon, and Ehren Pflugfelder, 1–5. New York: Association for Computing Machinery.

Durá, Lucía, Arvind Singhal, and Eliana Elias. 2013. "Minga Perú's Strategy for Social Change in the Perúvian Amazon: A Rhetorical Model for Participatory, Intercultural Practice to Advance Human Rights." *Rhetoric, Professional Communication, and Globalization* 4 (1): 33–54.

Evia, Carlos, and Ashley Patriarca. 2012. "Beyond Compliance: Participatory Translation of Safety Communication for Latino Construction Workers." *Journal of Business and Technical Communication* 26 (3): 340–67.

Flores, Nelson, and Jonathan Rosa. 2015. "Undoing Appropriateness: Raciolinguistic Ideologies and Language Diversity in Education." *Harvard Educational Review* 85 (2): 149–71.

Flores-Hutson, Patricia, Maria Isela Maier, Elvira Carrizal-Dukes, Lucía Durá, and Laura Gonzales. 2019. "La Salud en mis Manos: Localizing Health and Wellness Literacies in Transnational Communities through Participatory Mindfulness and Art-Based Projects." *Present Tense: A Journal of Rhetoric in Society* 3 (7). https://www.presenttensejournal.org/volume-7/la-salud-en-mis-manos/.

Frost, Erin. 2018. "Apparent Feminism and Risk Communication: Hazard, Outrage, Environment, and Embodiment." In *Key Theoretical Frameworks: Teaching Technical Communication in the Twenty-First Century*, edited by Angela M. Haas and Michelle F. Eble, 23–45. Boulder: University Press of Colorado.

Galván, Roberto A., and Richard V. Teschner, eds. 1977. *Dictionary of Chicano Spanish.* Vol. 5001. New York: National Textbook Company Trade.

Garba, Tapji, and Sara-Maria Sorentino. 2020. "Slavery Is a Metaphor: A Critical Commentary on Eve Tuck and K. Wayne Yang's 'Decolonization Is Not a Metaphor.' " *Antipode* 52 (3): 764–82.

García, Abigail Castellanos, Laura Gonzales, Cristina V. Kleinert, Tómas López Sarabia, Edith Matías Juan, Mónica Morales-Good, and Nora K. Rivera. 2022. *Indigenous Language Interpreters and Translators: Toward the Full Enactment of All Language Rights* Enculturation Intermezzo. https://intermezzo.enculturation.net/16-gonzales-et-al.htm.

García, Alma M. 1989. "The Development of Chicana Feminist Discourse, 1970–1980." *Gender and Society* 3 (2): 217–38.

García, Ofelia. 2009. "Education, Multilingualism, and Translanguaging in the 21st Century." *Social Justice through Multilingual Education* 143: 140–58.

García, Ofelia, and Li Wei. 2015. "Translanguaging, Bilingualism, and Bilingual Education." In *Handbook of Bilingual and Multilingual Education*, edited by Wayne E. Wright, Sovicheth Boun, and Ofelia García, 223–40. Hoboken, NJ: Wiley Blackwell.

Gómez-Barris, Macarena. 2017. *The Extractive Zone: Social Ecologies and Decolonial Perspectives.* Durham, NC: Duke University Press.

Gómez Mentjívar, Jennifer, and Gloria Elizabeth Chacón. 2019. *Indigenous Interfaces: Spaces, Technology, and Social Networks in Mexico and Central America.* Tucson: University of Arizona Press.

Gonzales, Laura. 2016. "(Re)Framing Multilingual Technical Communication with Indigenous Language Interpreters and Translators." *Technical Communication Quarterly* 31 (1): 1–16.

Gonzales, Laura. 2018. *Sites of Translation: What Multilinguals Can Teach Us about Digital Writing and Rhetoric.* Ann Arbor: University of Michigan Press.

Gonzales, Laura, and Rachel Bloom-Pojar. 2018. "A Dialogue with Medical Interpreters about Rhetoric, Culture, and Language." *Rhetoric of Health and Medicine* 1 (1): 193–212.

Gonzales, Laura, and Janine Butler. 2020. "Working toward Social Justice through Multilingualism, Multimodality, and Accessibility in Writing Classrooms." *Composition Forum* 44. https://compositionforum.com/issue/44/multilingualism.php.

Gonzales, Laura, and Mónica González Ybarra. 2020. "Multimodal Cuentos as Fugitive Literacies on the Mexico-US Borderlands." *English Education* 52 (3): 223–55.

GooglePlay. 2020. Limbu(kirati) Keyboard Plugin. https://play.google.com/store/apps/details?id=klye.plugin.lif&hl=en_US.

Grabill, Jeffrey T. 2007. *Writing Community Change: Designing Technologies for Citizen Action.* Cresshill, NJ: Hampton Press, Inc.

Grabill, Jeffrey T., and W. Michele Simmons. 1998. "Toward a Critical Rhetoric of Risk Communication: Producing Citizens and the Role of Technical Communicators." *Technical Communication Quarterly* 7 (4): 415–41.

Gubala, Carolyn, Kara Larson, and Lisa Melonçon. 2020. "Do Writing Errors Bother Professionals? An Analysis of the Most Bothersome Errors and How the Writer's Ethos Is Affected." *Journal of Business and Technical Communication* 34 (3): 250–86.

Haas, Angela M. 2007. "Wampum as Hypertext: An American Indian Intellectual Tradition of Multimedia Theory and Practice." *Studies in American Indian Literatures* 19 (4): 77–100.

Haas, Angela M. 2012. "Race, Rhetoric, and Technology: A Case Study of Decolonial Technical Communication Theory, Methodology, and Pedagogy." *Journal of Business and Technical Communication* 26 (3): 277–310.

Haas, Angela M. 2015. "Toward a Decolonial Digital and Visual American Indian Rhetorics Pedagogy." In *Survivance, Sovereignty, and Story: Teaching American Indian Rhetorics*, edited by Lisa King, Rose Gubele, and Joyce Rain Anderson, 188–208. Boulder: University Press of Colorado.

Haas, Angela M., and Michelle F. Eble, eds. 2018. *Key Theoretical Frameworks: Teaching Technical Communication in the Twenty-First Century*. Boulder: University Press of Colorado.

Hart-Davidson, William. 2001. "On Writing, Technical Communication, and Information Technology: The Core Competencies of Technical Communication." *Technical Communication* 48 (2): 145–55.

Hernández, Lorena Córdova. 2019. *Metáforas Ecológicas, Ideologías u Políticas Lingüísticas en la Revitalización de Lenguas Indígenas*. Oaxaca de Juárez, Mexico: Universidad Autónoma "Benito Juárez" de Oaxaca.

Hidalgo, Margarita. 1986. "Language Contact, Language Loyalty, and Language Prejudice on the Mexican Border." *Language in Society* 15 (2): 193–220.

Hitt, Allison. 2018. "Foregrounding Accessibility through (Inclusive) Universal Design in Professional Communication Curricula." *Business and Professional Communication Quarterly* 81 (1): 52–65.

hooks, bell. 1994. *Outlaw Culture: Resisting Representations*. New York: Routledge.

Hopton, Sarah Beth, and Rebecca Walton. 2019. "One Word of Heart Is Worth Three of Talent: Professional Communication Strategies in a Vietnamese Nonprofit Organization." *Technical Communication Quarterly* 28 (1): 39–53.

Hossain, Md. Faruk. 2020. "UX in Southeast Asia: Examples across Current UX Maturity Levels." *User Experience Magazine*. http://uxpamagazine.org/ux-in-southeast-asia-examples-across-current-ux-maturity-levels/.

Hubrig, Ada, Ruth Osorio, Neil Simpkins, Leslie R. Anglesey, Ellen Cecil-Lemkin, Margaret Fink, Janine Butler, Tonya Stremlau, Stephanie L. Kerschbaum, Brenda Jo Brueggemann, et al. 2020. "Enacting a Culture of Access in Our Conference Spaces." *College Composition and Communication* 72 (1): 87–117.

Hull, Brittany, Cecilia D. Shelton, and Temptaous Mckoy. 2020. "Dressed by Not Tryin' to Impress: Black Women Deconstructing 'Professional' Dress." *Journal of Multimodal Rhetorics* 3 (2): 7–20.

Itchuaqiyaq, Cana. 2021. "Iñupiat Iḷitqusiat: An Indigenist Ethics Approach for Working with Marginalized Knowledges in Technical Communication." In *Equipping Technical Communicators for Social Justice Work: Theories, Methodologies, and Topics*, edited by Rebecca Walton and Godwin Agboka, 33–48. Logan: Utah State University Press.

Jones, Natasha N. 2016. "The Technical Communicator as Advocate: Integrating a Social Justice Approach in Technical Communication." *Journal of Technical Writing and Communication* 46 (3): 342–61.

Jones, Natasha N. 2020. "Coalitional Learning in the Contact Zones: Inclusion and Narrative Inquiry in Technical Communication and Composition Studies." *College English* 82 (5): 515–26.

Jones, Natasha N., Kristen R. Moore, and Rebecca Walton. 2016. "Disrupting the Past to Disrupt the Future: An Antenarrative of Technical Communication." *Technical Communication Quarterly* 25 (4): 211–29.

Jones, Natasha N., and Miriam F. Williams. 2017. "The Social Justice Impact of Plain Language: A Critical Approach to Plain-Language Analysis." *IEEE Transactions on Professional Communication* 60 (4): 412–29.

Jones, Natasha N., and Miriam F. Williams. 2018. "Technologies of Disenfranchisement: Literacy Tests and Black Voters in the US from 1890 to 1965." *Technical Communication* 65 (4): 371–86.

Jung, Julie. 2014. "Interdependency as an Ethic for Accessible Intellectual Publics." *Reflections: A Journal of Community-Engaged Writing and Rhetoric* 14 (1): 101–20.

Kerschbaum, Stephanie L. 2014. "Toward a New Rhetoric of Difference." Urbana, IL: Conference on College Composition and Communication, National Council of Teachers of English.

Kleinert, Cristina, and Christiane Stallaert. 2015. "La Formación de Intérpretes de Lenguas Indígenas para la Justicia en México: Sociología de las Ausencias y Agencia Decolonial." *Sendebar* 26: 235–54.

Lara, Clara Castillo. 2017. "La Constitución Mexicana y el Convenio 169 de la OIT Sobre Pueblos Indígenas y Tribales." *Alegatos* 97: 559–78.

Liasidou, Anastasia. 2013. "Intersectional Understandings of Disability and Implications for a Social Justice Reform Agenda in Education Policy and Practice." *Disability and Society* 28 (3): 299–312.

Loebick, Karla, and J. Estrella Torrez. 2016. "Where You Are from Defines You: Intersection of Community Engagement, Border Pedagogy, and Higher Education." *Journal of Public Scholarship in Higher Education* 6: 21–44.

Lorde, Audre. 2012. *Sister Outsider: Essays and Speeches.* Berkeley, CA: Crossing Press.

MacGregor-Mendoza, Patricia. 1999. *Spanish and Academic Achievement among Midwest Mexican Youth: The Myth of the Barrier.* New York: Garland.

Makoni, Sinfree, and Alastair Pennycook, eds. 2006. *Disinventing and Reconstituting Languages.* Vol. 62. Bristol, UK: Multilingual Matters.

Martínez, Oscar Jáquez. 1994. *Border People: Life and Society in the US-Mexico Borderlands.* Tucson: University of Arizona Press.

Maylath, Bruce, and Kirk St.Amant. 2019. *Translation and Localization: A Guide for Technical and Professional Communicators.* New York: Routledge.

Medina, Cruz. 2014. *Reclaiming Poch@ Pop: Examining the Rhetoric of Cultural Deficiency.* New York: Springer.

Medina, Jennifer. 2019. "Anyone Speak K'iche or Mam? Immigration Courts Overwhelmed by Indigenous Languages." *New York Times,* May 19, 2019. https://www.nytimes.com/2019/03/19/us/translators-border-wall-immigration.html.

Mignolo, Walter. 2000. "The Many Faces of Cosmo-Polis: Border Thinking and Critical Cosmopolitanism." *Public Culture* 12 (3): 721–48.

Milu, Esther. 2021. "Diversity of Raciolinguistic Experiences in the Writing Classroom: An Argument for a Transnational Black Language Pedagogy." *College English* 83 (6): 415–41.

Ministry of Culture, Tourism, and Civil Aviation. 2020. *Nepal Tourism Statistics 2019.* Singhadurbar, Kathmandu: Government of Nepal. https://www.tourism.gov.np/files/NOTICE%20MANAGER_FILES/Nepal_%20tourism_statics_2019.pdf.

Morales-Good, Mónica. 2022. "Roundtable on Translator Training and Professionalization." In *Indigenous Language Interpreters and Translators: Toward the Full Enactment of all Language Rights,* by Abigail Castellanos García, Laura Gonzales, Cristiva V. Kleinert, Tomás López Sarabia, Edith Matías Juan, Mónica Morales-Good, Nora K. Rivera. Lexington, KY: *Intermezzo.* Web. https://intermezzo.enculturation.net/16-gonzales-et-al.htm.

Nepali Unicode Converter. 2020. https://www.nepali-unicode.com.

Noble, Safiya Umoja. 2018. *Algorithms of Oppression: How Search Engines Reinforce Racism.* New York: New York University Press.
Nuñez, Idalia. 2019. "'Le Hacemos La Lucha': Learning from Madres Mexicanas' Multimodal Approaches to Raising Bilingual, Biliterate Children." *Language Arts* 97 (1): 7–16.
Omniglot: The Online Encyclopedia of Writing Systems and Languages. 2020. "Kirat Rai Script." https://omniglot.com/writing/bantawa.htm.
Pandey, Shyam B. 2020. "English in Nepal: A Sociolinguistic Profile." *World Englishes* 39 (3): 1–14.
Peñalosa, Fernando. 1980. *Chicano Sociolinguistics: A Brief Introduction.* Rowley, MA: Newbury House Publishers.
Pérez-Quiñones, Manuel, and Consuelo Carr Salas. 2021. "How the Ideology of Monolingualism Drives Us to Monolingual Interaction." *Interactions* 28 (3): 66–69.
Perkins, Jane, and Nancy Blyler. 1999. *Narrative and Professional Communication.* Stamford, CT: Ablex.
Piepzna-Samarasinha, Leah Lakshmi. 2018. *Care Work: Dreaming Disability Justice.* Vancouver: Arsenal Pulp Press.
Pimentel, Octavio. 2008. "Disrupting Discourse: Introducing Mexicano Immigrant Success Stories." *Reflections: A Journal of Community-Engaged Writing and Rhetoric* 8: 171–96.
Poudyal, Bibhushana. 2018. "Building Critical Decolonial Digital Archives: Recognizing Complexities to Reimagine Possibilities." *Xchanges: An Interdisciplinary Journal of Technical Communication, Rhetoric, and Writing across the Curriculum* 13 (2): 1–16.
Poudyal, Bibhushana, and Laura Gonzales. 2019. "'So You Want to Build a Digital Archive?' A Dialogue on Digital Humanities Graduate Pedagogy." *Journal of Interactive Technology and Pedagogy* 15. https://jitp.commons.gc.cuny.edu/page/2/?s=first+day.
Prajapati, Chetan, Jwalanta Deep Shrestha, and Shishir Jha. 2008. *Nepali Unicode Keyboard Layout Standarization Based on Genetic Algorithm.* http://www.jwalanta.com.np/nepalikeyboard.
Price, Margaret. 2011. *Mad at School: Rhetorics of Mental Disability and Academic Life.* Ann Arbor: University of Michigan Press.
Price, Margaret, and Stephanie L. Kerschbaum. 2016. "Stories of Methodology: Interviewing Sideways, Crooked, and Crip." *Canadian Journal of Disability Studies* 5 (3): 18–56.
Racadio, Robert, Emma J. Rose, and Beth E. Kolko. 2014. "Research at the Margin: Participatory Design and Community Based Participatory Research." In *Proceedings of the 13th Participatory Design Conference: Short Papers, Industry Cases, Workshop Descriptions, Doctoral Consortium Papers, and Keynote Abstracts,* edited by Heike Winschiers-Theophilus, Vicenzo D'Andrea, and Ole Sejer Iversen, 49–52. New York: Association for Computing Machinery.
Redish, Janice, and Carol Barnum. 2011. "Overlap, Influence, Intertwining: The Interplay of UX and Technical Communication." *Journal of Usability Studies* 6 (3): 90–101.
Ríos, Gabriela R. 2015. "Cultivating Land-Based Literacies and Rhetorics." *Literacy in Composition Studies* 3 (1): 60–70.
Rosa, Jonathan, and Nelson Flores. 2017. "Unsettling Race and Language: Toward a Raciolinguistic Perspective." *Language in Society* 46 (5): 621–47.
Rose, Emma, Elin Björling, and Maya Cakmak. 2019. "Participatory Design with Teens: A Social Robot Design Challenge." In *Proceedings of the 18th ACM International Conference on Interaction Design and Children,* edited by Jerry Alan Fails, 604–9. New York: Association for Computing Machinery.
Rose, Emma, and Alison Cardinal. 2018. "Participatory Video Methods in UX: Sharing Power with Users to Gain Insights into Everyday Life." *Communication Design Quarterly* 6 (2): 9–20.
Rose, Emma, and Alison Cardinal. 2021. "Purpose and Participation: Heuristics for Planning, Implementing, and Reflecting on Social Justice Work." In *Equipping Technical*

*Communicators for Social Justice Work: Theories, Methodologies, and Topics*, edited by Rebecca Walton and Godwin Agboka, 75–97. Logan: Utah State University Press.

Rose, Emma J., Robert Racadio, Kalen Wong, Shally Nguyen, Jee Kim, and Abbie Zahler. 2017. "Community-Based User Experience: Evaluating the Usability of Health Insurance Information with Immigrant Patients." *IEEE Transactions on Professional Communication* 60 (2): 214–31.

Sackey, Donnie Johnson. 2020. "One-Size-Fits-None: A Heuristic for Proactive Value Sensitive Environmental Design." *Technical Communication Quarterly* 29 (1): 33–48.

Sandoval, Chela. 2013. *Methodology of the Oppressed.* Vol. 18. Minneapolis: University of Minnesota Press.

San Pedro, Timothy, and Valerie Kinloch. 2017. "Toward Projects in Humanization: Research on Co-creating and Sustaining Dialogic Relationships." *American Educational Research Journal* 54 (1): 373–94.

Schalk, Sami. 2018. *Bodyminds Reimagined: (Dis)ability, Race, and Gender in Black Women's Speculative Fiction.* Durham, NC: Duke University Press.

Shannon, Claude, and Warren Weaver. 1949. *The Mathematical Theory of Communication.* Urbana: University of Illinois Press.

Sharma, Bal Krishna, and Prem Phyak. 2017. "Neoliberalism, Linguistic Commodification, and Ethnolinguistic Identity in Multilingual Nepal." *Language in Society* 46 (2): 231–56.

Shivers-McNair, Ann, and Clarissa San Diego. 2017. "Localizing Communities, Goals, Communication, and Inclusion: A Collaborative Approach." *Technical Communication* 64 (2): 97–112.

Shrestha, Sagun. 2016. "Role and Status of English and Other Languages in Nepal." *Journal of NELTA* 21 (1–2): 105–12.

Simmons, W. Michele. 2008. *Participation and Power: Civic Discourse in Environmental Policy Decisions.* Albany: State University of New York Press.

Simpson, Leanne Betasamosake. 2017. *As We Have Always Done: Indigenous Freedom through Radical Resistance.* Minneapolis: University of Minnesota Press.

Slack, Jennifer Daryl, David James Miller, and Jeffrey Doak. 1993. "The Technical Communicator as Author: Meaning, Power, Authority." *Journal of Business and Technical Communication* 7 (1): 12–36.

Smith, Linda Tuhiwai. 1999. *Decolonizing Methodologies: Research and Indigenous Peoples.* London: Zed Books Ltd.

South Asian Foundation for Academic Research. 2020. "Misssion." http://www.safarsouthasia.org/mission/.

Sun, Huatong. 2012. *Cross-Cultural Technology Design: Creating Culture-Sensitive Technology for Local Users.* New York: Oxford University Press.

Swain, Elise. 2019, November 15. "The Coup That Ousted Bolivia's Evo Morales Is Another Setback for Latin American Socialism." *The Intercept.* https://theintercept.com/2019/11/15/bolivia-evo-morales-coup-brazil-intercepted/.

Teston, Christa, Laura Gonzales, Kristin Marie Bivens, and Kelly Whitney. 2019. "Surveying Precarious Publics." *Rhetoric of Health and Medicine* 2 (3): 321–51.

Texas Office of Court Administration. 2019. "Licensed Court Interpreters." https://www.txcourts.gov/jbcc/licensed-court-interpreters.aspx.

Tlostanova, Madina V., and Walter D. Mignolo. 2012. *Learning to Unlearn: Decolonial Reflections from Eurasia and the Americas.* Columbus: Ohio State University Press.

Torrez, J. Estrella, Laura Gonzales, Victor J. Del Hierro, Santos Ramos, and Everardo Cuevas. 2019. "Comunidad de Cuentistas: Making Space for Indigenous and Latinx Storytellers." *English Journal* 108 (3): 44–50.

Tuck, Eve. 2009. "Suspending Damage: A Letter to Communities." *Harvard Educational Review* 79 (3): 409–28.

Tuck, Eve, and K. Wayne Yang. 2012. "Decolonization Is Not a Metaphor." *Decolonization: Indigeneity, Education, and Society* 1 (1): 1–40.

Unconference. Net. 2020. "Unconference: Like a Conference, Only Better." http://unconference.net/methods/.

UNESCO. 2019, January 25. "Presentación del Año Internacional de las Lenguas Indígenas 2019." UNESCO. https://es.unesco.org/news/presentacion-del-ano-internacional lenguas-indigenas-2019.

US Embassy in Nepal. 2018. "Number of Nepali Students Reach[es] Highest in the Last Three Years." https://np.usembassy.gov/number-of-nepali-students-reach-highest-in-the-last-three-years/.

Vatrapu, Ravi, and Manuel A. Pérez-Quiñones. 2006. "Culture and Usability Evaluation: The Effects of Culture in Structured Interviews." *Journal of Usability Studies* 1 (4): 156–70.

Villamil, Alba. 2020. "How We Empathize in UX Matters." *dscout.* https://dscout.com/people-nerds/how-we-empathize-in-ux.

Walton, Rebecca. 2016. "Supporting Human Dignity and Human Rights: A Call to Adopt the First Principle of Human-Centered Design." *Journal of Technical Writing and Communication* 46 (4): 402–26.

Walton, Rebecca, and Sarah Beth Hopton. 2018. " 'All Vietnamese Men Are Brothers': Rhetorical Strategies and Community Engagement Practices Used to Support Victims of Agent Orange." *Technical Communication* 65 (3): 309–25.

Walton, Rebecca, Kristen Moore, and Natasha Jones. 2019. *Technical Communication after the Social Justice Turn: Building Coalitions for Action.* New York: Routledge.

Walton, Rebecca, Maggie Zraly, and Jean Pierre Mugengana. 2015. "Values and Validity: Navigating Messiness in a Community-Based Research Project in Rwanda." *Technical Communication Quarterly* 24 (1): 45–69.

Walwema, Josephine. 2016. "Tailoring Information and Communication Design to Diverse International and Intercultural Audiences: How Culturally Sensitive ICD Improves Online Market Penetration." *Technical Communication* 63 (1): 38–52.

Walwema, Josephine. 2021. "Rhetoric and Cape Town's Campaign to Defeat Day Zero." *Journal of Technical Writing and Communication* 51 (2): 103–36.

Williams, Miriam F. 2013. "A Survey of Emerging Research: Debunking the Fallacy of Colorblind Technical Communication." *Programmatic Perspectives* 5 (1): 86–93.

Williams, Miriam F., and Octavio Pimentel, eds. 2014. *Communicating Race, Ethnicity, and Identity in Technical Communication.* Amityville, NY: Baywood.

Winschiers, Heike, and Jens Fendler. 2007. "Assumptions Considered Harmful." In *International Conference on Usability and Internationalization,* edited by Nuray Aykin, 452–61. Berlin: Springer: Berlin, Heidelberg.

Yergeau, Melanie, Elizabeth Brewer, Stephanie L. Kerschbaum, Sushil Oswal, Margaret Price, Michael J. Salvo, Cynthia L. Selfe, and Franny Howes. 2013. "Multimodality in Motion: Disability and Kairotic Spaces." *Kairos* 18 (1). http://kairos.technorhetoric.net/18.1/coverweb/yergeau-et-al/.

Zdenek, Sean. 2015. *Reading Sounds: Closed-Captioned Media and Popular Culture.* Chicago: University of Chicago Press.

# INDEX